重庆市骨干高等职业院校建设项目规划教材
重庆水利电力职业技术学院课程改革系列教材

水工建筑行业英语

主　编　何德胜　孙晓玲　朱小华
副主编　阳　林　曾　英　姚嫚珍
　　　　王　宁
主　审　费文平

黄河水利出版社
·郑州·

内 容 提 要

本书是重庆市骨干高等职业院校建设项目规划教材、重庆水利电力职业技术学院课程改革系列教材之一,由重庆市财政重点支持,根据高职高专教育水工建筑行业英语课程标准及理实一体化教学要求编写完成。本书以职业岗位对人才的要求以及学生未来职业发展的要求为出发点,共 8 个单元,涵盖水利学科主要分支。每个单元设计有不同的模块和练习,围绕读、译、说、写、听等综合能力训练目标展开,以培养学生在将来工作中所需要的英语应用能力,在帮助学生打好语言基础的同时,特别强调在行业工作过程中的英语交际能力。

本书可供高职高专院校水利水电建筑工程专业教学使用,也可供相关专业工程技术人员学习参考。

图书在版编目(CIP)数据

水工建筑行业英语/何德胜,孙晓玲,朱小华主编.—郑州:黄河水利出版社,2017.6
重庆市骨干高等职业院校建设项目规划教材
ISBN 978–7–5509–1589–3

Ⅰ.①水… Ⅱ.①何… ②孙…③朱… Ⅲ.①水工建筑物–英语–高等职业教育–教材 Ⅳ.①TV6

中国版本图书馆 CIP 数据核字(2016)第 302753 号

组稿编辑:王路平 电话:0371-66022212 E-mail:hhslwlp@163.com

| 出 版 社:黄河水利出版社 | 网址:www.yrcp.com |
| 地址:河南省郑州市顺河路黄委会综合楼 14 层 | 邮政编码:450003 |

发行单位:黄河水利出版社
  发行部电话:0371-66026940、66020550、66028024、66022620(传真)
  E-mail:hhslcbs@126.com
承印单位:虎彩印艺股份有限公司
开本:787 mm×1 092 mm 1/16
印张:6.75
字数:200 千字
版次:2017 年 6 月第 1 版 印次:2017 年 6 月第 1 次印刷

定价:20.00 元

序

近年来,我国高职教育改革方兴未艾,遗憾的是,高职英语教学远远没有跟上改革的步伐,以大学英语A级、B级与三级、四级考试为核心的教学目标常年没有变化,纯粹的知识"填鸭"式教育使得外语基础本来就偏弱的高职学生对学习提不起兴趣,更为重要的是,与其他专业课程相比,花费大量时间与精力的英语教学与高职人才培养目标——厚基础、宽专业、强能力、高素质渐行渐远。面对尴尬的现实,人们不禁要问:高职学生学习英语的目的是什么?理论与现实的不断反思告诉我们:灵活运用语言工具为未来职业的可持续发展服务才是当前高职学生学习英语的切实目标。我们培养的是面向生产、建设、服务和管理等第一线需要的高技能人才,而不是高科技、科研型人才。以此目标为导向,在高职高专英语课程设置上,加开专业性较强的行业英语是十分必要的。高职学生在学习、工作中最为常见的外语需求就是看得懂、说得出、写得了有关外语资料,特别是结合专业和岗位得体地进行日常会话和专业交流即可。

职业院校的定位决定了其所开设的课程应符合"以服务为宗旨、以就业为导向"的办学方针,凸显职业教育的特色。在变革的大趋势下,《高等职业教育英语课程教学要求》将行业英语教学纳入了高职高专公共英语教学的内容,许多高职院校也改变了过去没有明确职业指向的通用英语(EGP)教学状况,陆续开展了专门用途英语(ESP)教学改革。但困扰很多高职院校英语教师的问题并没有得到很好的解决。比如,高职院校行业英语的含义到底该如何理解?行业英语和专业英语的区别在哪里?高职院校的英语教学需要在多大程度上与行业结合?该如何与行业相结合?

理清这些基本概念将有助于我们理清教学改革的思路,坚定改革必胜的信念。针对高职院校教学中行业英语的性质,斯特雷文斯(Strevens)提出了ESP的四个区别性特征:①需求上满足特定的学习者;②内容上与特定的专业和职业相关;③词汇、句法和语篇学习以特定专业和职业相关的活动中的语言运用为主;④与普通英语形成对照。

几年前,在英语教材市场,没有专门的国家统编的高职英语教材,因此高职院校不得不沿用本科教材。目前,仍有部分高职院校在使用本科英语教材。本科教育担负着为社会培养学术型和工程型高级专门人才的重任,职业教育则担负着为社会培养高级应用型技术人才的任务。两者的人才培养目标不同,教材自然有区别。高职教材与普通高校教材的主要区别在于:高职教材强调"职业性",而普通高校教材更强调"学术性";高职教材比普通高校教材更强调"应用性、实用性";高职专业课教材强调"技能性",而普通高校同类教材更强调"理论研究";高职教材比普通高校教材更强调"开放性""发展性",及时补充新知识和新技能;高职教材更新速度快,而普通高校教材则相对稳定;生源的多样性特点,使得高职教材比普通高校教材更强调"多样性"。

目前尚没有一套完整的符合职业技术学院使用的英语教材,主要有以下几个方面的原因：

（1）有的教材虽实用性较强，注重了"交际原则"，但忽略了"系统原则"，没有遵循"循序渐进原则"或"阶段性原则"。

（2）有些教材邀请了几十所学校参编，每所学校编一个单元，再拼凑成一本书，其连贯性差、跳跃度大，给教学带来了困难，因此不少学校不得不中断使用。

（3）有些教材起点偏高，选材多注重自然科学和文学作品，不适应学生入学时的英语水平和就业后的应用需要。

（4）有些教材过分偏重某一个方面能力的训练，等等。

目前，高职英语使用校本教材的现象也很普遍。首先，这是因为随着科技的发展，经济的全球化、多元化，职业变化加速，高职院校应紧扣市场经济的发展，利用校本资源，开发英语校本教材；其次，我国的高职教育还处于初级阶段，正是高职教育生存与发展的关键时期，能否健康地开展取决于高职院校自身的发展和特色，校本教材的开发则是特色学校的必由之路；再次，高职院校学生来源复杂，有职高生、普高生、初中毕业生、社会待业青年等，学生群体的多样性造成依靠国家课程和地方课程难以满足其身心健康发展的要求，而且生源的多样性使得高职学生英语基础在所有学科中最为参差不齐，因此高职院校必须根据自身的发展定位，进行有效的资源整合，开发英语校本教材，打造出有特色的高职毕业生。

重庆水利电力职业技术学院
《水工建筑行业英语》教材编写组
2016年10月

前　言

按照"重庆市骨干高等职业院校建设项目"规划要求,水利水电建筑工程专业是该项目的重点建设专业之一,由重庆市财政支持、重庆水利电力职业技术学院负责组织实施。按照子项目建设方案和任务书,通过广泛地深入行业、市场调研,与行业、企业专家共同研讨,不断创新基于职业岗位能力的"三轮递进,两线融通"的人才培养模式,以水利水电建设一线的主要技术岗位核心能力为主线,兼顾学生职业迁徙和可持续发展需要,构建基于职业岗位能力分析的教学做一体化课程体系,优化课程内容,进行精品资源共享课程与优质核心课程的建设。经过3年的探索和实践,已形成初步建设成果。为了固化骨干建设成果,进一步将其应用到教学之中,最终实现让学生受益,经学院审核,决定正式出版系列课程改革教材,包括优质核心课程和精品资源共享课程等。

当前,世界科学技术发展十分迅速,为了了解、学习和借鉴国外先进的科学技术,为我国的社会主义建设服务,需要大量地阅读和翻译国外科技文献资料。另外,近年来和在以后的若干年内,我国在水利水电建设中,从国外引进了且还将继续引进大批先进技术和设备,为了尽快消化这些新技术和装好、用好及管理好这些设备,也需要详细地阅读和翻译引进技术和设备的技术说明文件。要提高专业科技英语的阅读和翻译能力,除需要掌握英语语法的基本知识和基本词汇及具有相当广泛的专业知识外,还必须熟悉专业词汇和科技英语中一些常用词、词组或短语,熟悉科技英语中常见的句型和文体,以及掌握翻译科技文献的基本技巧。

本课程立足于水工建筑专业群所面向的行业,依据典型工作环节或场景设计教学内容,以任务为驱动开展语言教学,力求使学生具备在本行业领域内运用英语进行基本交流的能力:①熟练掌握水工建筑施工过程中的常用交际语,如自我介绍、模拟不同国籍的人就工作职务和技术专业现场会话、水工建筑工地会话、在建筑工地用英语交流天气与环境对施工的影响等;②了解并掌握水工建筑行业英语一些常见的专业词汇和科普文献的阅读、翻译,如水分循环及径流形成,水文资料的收集,水文统计的基本方法,年径流和多年平均年输沙量的计算,岩基上的重力坝,土石坝与堤防,水电站水力过渡过程,水电站厂房、水利水电枢纽布置,农田水分状况和灌溉用水量,地面灌水技术,灌溉水源及取引水工程技术,灌溉渠道系统规划设计,喷灌工程技术,低压管道输水灌溉工程技术等;③了解并掌握水工建筑行业英语的一些常见应用文写作,如工程的招投标,合同的谈判和签订,工程材料的询价、报价等。

本教材由重庆水利电力职业技术学院组织编写,由何德胜、孙晓玲、朱小华担任主编,阳林、曾英、姚嫚珍、王宁担任副主编,由四川大学水电学院费文平教授担任主审。特别感谢费文平教授对本教材的编写所付出的辛苦和努力!

由于该教材的编写尚在起步中,不足之处在所难免,真诚欢迎各位读者及同仁赐教!

编　者
2016 年 8 月

目 录

序
前言

UNIT ONE　Planning of hydrology and water resources and hydropower
单元一　水文与水利水电规划 ……………………………………………………（1）
　　Speaking ………………………………………………………………………（1）
　　Text 1　Planning of hydrology and water resources and hydropower ………（3）
　　主题1　水文与水利水电规划 ………………………………………………（5）
　　Writing …………………………………………………………………………（6）
　　专业英语中常用的词头 ………………………………………………………（7）

UNIT TWO　Hydraulic structures
单元二　水利水电工程建筑物 …………………………………………………（11）
　　Speaking ………………………………………………………………………（11）
　　Text 2　Hydraulic structures ………………………………………………（12）
　　主题2　水利水电工程建筑物 ………………………………………………（14）
　　Writing …………………………………………………………………………（16）
　　词义的处理 ……………………………………………………………………（18）

UNIT THREE　Irrigation and drainage engineering
单元三　灌溉排水工程技术 ……………………………………………………（21）
　　Speaking ………………………………………………………………………（21）
　　Text 3　Irrigation and drainage engineering ………………………………（24）
　　主题3　灌溉排水工程技术 …………………………………………………（27）
　　Writing …………………………………………………………………………（30）
　　科技术语翻译 …………………………………………………………………（32）

UNIT FOUR　Construction organization and management of hydraulic projects
单元四　水利水电工程施工组织与管理 ………………………………………（34）
　　Speaking ………………………………………………………………………（34）
　　Text 4　Construction organization and management of hydraulic projects ………（33）
　　主题4　水利水电工程施工组织与管理 ……………………………………（41）
　　Writing …………………………………………………………………………（43）
　　被动语态的几种表示形式 ……………………………………………………（44）
　　被动语态的翻译 ………………………………………………………………（47）

UNIT FIVE　Hydraulic engineering management
单元五　水利工程管理 …………………………………………………………………… (50)
- Speaking ………………………………………………………………………………… (50)
- Text 5　Hydraulic engineering management ………………………………………… (51)
- 主题 5　水利工程管理 ………………………………………………………………… (55)
- Writing …………………………………………………………………………………… (57)
- 否定结构 ………………………………………………………………………………… (58)

UNIT SIX　Construction techniques of hydraulic projects
单元六　水利水电工程施工技术 …………………………………………………………… (62)
- Speaking ………………………………………………………………………………… (62)
- Text 6　Construction techniques of hydraulic projects ……………………………… (64)
- 主题 6　水利水电工程施工技术 ……………………………………………………… (68)
- Writing …………………………………………………………………………………… (69)
- 否定结构 ………………………………………………………………………………… (71)

UNIT SEVEN　Drainage system in our country
单元七　我国水系 …………………………………………………………………………… (76)
- Speaking ………………………………………………………………………………… (76)
- Text 7　Drainage system in our country ……………………………………………… (77)
- 主题 7　我国水系 ……………………………………………………………………… (78)
- Writing …………………………………………………………………………………… (79)
- 同位语和插入语 ………………………………………………………………………… (81)

UNIT EIGHT　Review test
单元八　复习测试 …………………………………………………………………………… (85)
- Speaking ………………………………………………………………………………… (85)
- Writing …………………………………………………………………………………… (87)
- 数字大小及数字增减的表示方法与翻译方法 ………………………………………… (88)
- As 的主要用法及翻译 ………………………………………………………………… (91)

参考文献 …………………………………………………………………………………… (99)

UNIT ONE Planning of hydrology and water resources and hydropower

水文与水利水电规划

 Speaking

 Welcome and introduction 问候和介绍

1. Welcome to our job site.

   欢迎到我们工地来。

2. I wish we shall have a friendly cooperation in coming days.

   希望今后友好合作。

3. Please allow me to introduce a fellow of mine, Mr. ____.

   请允许我给你介绍一位我的同事,____先生。

4. I am a manager(project manager).

   我是经理(项目经理)。

5. My technical specialty is civil engineering (mechanical equipment, electrical).

   我的技术专业是土建工程(机械设备、电气)。

6. I am a mechanician(electrician, builder, erector, concrete worker).

   我是一名机械工(电工、建筑工人、安装工人、混凝土工)。

7. What is your nationality? Are you American?

   你是什么国籍的?你是美国人吗?

Introduction of hydroelectric engineering

The Three Gorges Dam

  The Three Gorges Project is also called the Three Gorges Dam, the Three Gorges Hydropower Station, located in Sandouping Town, Yichang City, Hubei Province, is the largest hydropower station in the world, also is the largest projects that have ever been built in China. The Three Gorges Dam is concrete gravity dam, the total length of 2 309.47 meters, the crest elevation of 185 meters, water elevation 175 meters, equipped with 32 single capacity of 700 000 kilowatts of hydropower unit, installed capacity of 22.4 million kilowatts. The Three Gorges reservoir is more than 600 kilometers long, with a total investment of 95.46 billion yuan.

  三峡工程又称三峡大坝、三峡水电站，位于湖北省宜昌市的三斗坪镇，是世界上规模最大的水电站，也是中国有史以来建设的最大型的工程项目。三峡大坝为混凝土重力坝，总长度2 309.47 m，坝顶高程185 m，蓄水高程175 m，安装有32台单机容量为70万kW的水电机组，装机容量达到2 240万kW。三峡水库长600多km，总投资954.6亿元人民币。

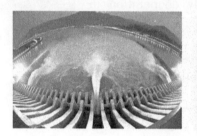

三峡大坝首次开闸泄洪

Text 1 Planning of hydrology and water resources and hydropower

Water is a kind of important natural resource, as well as the basic material conditions of existence and development for human and even the entire ecosystem. Water resource refers to the amount of fresh water available for exploitation by human under current economic and technological conditions, a dynamic value which can be recovered and updated in a certain period of time, and a kind of dynamic resource. Hydrological phenomenon is formed by circulating changes of various water resources in the nature, which is a result of the combined action of many complex factors. Planning of water resources and hydropower is to study how to develop and utilize water resources in a reasonable economic manner, to govern the rivers, to determine the ways, scale and efficiency of developing water resources and hydropower project, as well as to draw up a reasonable management and utilization mode of water resources and hydropower project, according to the actual needs of the national economy and the objective situation of water resources.

1.1 Water circulation and runoff formation

Phenomenon of water constantly alternating and transferring is known as the hydrological cycle, also known as the hydrologic cycle, water cycle for short. Water cycle includes rivers, precipitation, evaporation, infiltration, runoff, and other links. Water movement follows the law of conservation of mass.

1.2 Collection of hydrological data

Hydrological stations are necessary basic units to carry out hydrological observations and obtain hydrological data for analysis and calculation of hydrology and water resources. Hydrological information collected includes precipitation, surface evaporation, water level, discharge, sediment, and floods, low water and storm survey.

1.3 Basic methods of hydrological statistics

Statistical law of hydrological phenomena is sought through mathematical statistics called hydrological statistics in hydrology, which includes frequency calculation and correlation analysis.

1.4 Calculations of annual runoff and mean annual sediment

Analysis and calculation of annual runoff and sediment can provide services for planning and designing of hydroelectric engineering. Analysis and calculation results of annual runoff are used for regulation calculation of reservoir in coordination with water data. Calculation results of mean annual sediment can provide an important basis for the selection of level of dead water of reservoir.

1.5 Analysis and calculation of design flood

Design flood is the basis of designing of hydraulic structures, which consists of three elements: the design peak flow, total design flood and design flood hydrograph.

1.6 Calculation of reservoir profiting regulation

Runoff regulation is to build various hydraulic projects to control runoff of rivers for solving the contradiction between supply and demand, and to redistribute runoff in time and space according to the requirements from all kinds of water departments in order to solve the incompatible contradictions between water supply and water use in time and quantity. Runoff regulation includes profiting regulation and flood control regulation.

1.7 Hydropower calculation of hydropower station

Hydropower calculation is mainly to determine the energy indexes of the hydropower station: firm capacity and mean annual electricity generation, and corresponding main parameters: installed capacity and normal water level of reservoir.

1.8 Calculation of reservoir flood control regulation

Flood control regulation is to retain and impound part of floods temporarily in the reservoir when the floods into reservoir are great, and release them gradually after the peak flood passes, to ensure the safety of the dam and prevent downstream areas from flooding.

1.9 Reservoir scheduling

Reservoir scheduling is to make use of reservoir storage capacity in accordance with the requirements of the national economy sectors, to control and regulate natural runoff of rivers by various buildings and equipments of hydro-junction to achieve a goal of eliminating flood disasters and promoting water conservancy. It is also known as reservoir controlling and application.

主题1　水文与水利水电规划

　　水是一种重要的自然资源,也是人类乃至整个生态系统赖以存在和发展的基本物质条件。水资源是指在目前的经济和技术条件下,可供人们开发利用的淡水量,是在一定时间内可以得到恢复和更新的动态量,是一种动态资源。水文现象是由自然界中各种水体的循环变化所形成的,是许许多多复杂影响因素综合作用的结果。水利水电规划是根据国民经济的实际需要,以及水资源的客观情况,研究如何经济、合理地开发利用水资源、治理河流,确定水利水电工程的开发方式、规模和效益,以及拟定水利水电工程的合理管理运用方式等。

1.1　水分循环及径流形成

　　水分不断交替转移的现象称为水分循环,也叫水文循环,简称水循环。水循环包括河流、降水、蒸发、下渗、径流等环节。水分运动遵循质量守恒定律。

1.2　水文资料的收集

　　水文测站是进行水文观测、获取水文水资源分析计算所必须的水文资料的基层单位。采集的水文信息主要包括降水量、水面蒸发量、水位、流量、泥沙以及洪水、枯水和暴雨调查。

1.3　水文统计的基本方法

　　水文现象的统计规律通过数理统计来寻求,在水文学中称为水文统计,包括频率计算和相关分析。

1.4　年径流和多年平均年输沙量的计算

　　年径流及年输沙量的分析计算是为水利水电工程的规划设计服务的。年径流分析计算成果与用水资料相配合,进行水库调节计算;多年平均年输沙量计算成果为水库死水位的选择提供了重要依据。

1.5　设计洪水的分析计算

　　设计洪水是水工建筑物设计的依据,它包括三个要素:设计洪峰流量、设计洪水总量和设计洪水过程线。

1.6 水库兴利调节计算

径流调节是为解决供与需的矛盾而修建各种水利工程对河川径流进行控制,并按各用水部门的要求对其进行时间和空间上的重新分配,以解决来水与用水在时间上与数量上不相适应的矛盾。径流调节分为兴利调节和防洪调节。

1.7 水电站水能计算

水能计算主要是确定水电站的动能指标——保证出力和多年平均年发电量,及其相应的主要参数——装机容量和水库的正常蓄水位。

1.8 水库防洪调节计算

防洪调节是当入库洪水很大时,为确保水库大坝安全和下游地区免受洪灾,临时将部分洪水拦蓄在水库中,待到洪峰过后再陆续放出。

1.9 水库调度

水库调度是指按照国民经济各部门的要求,运用水库的调蓄能力,由水利枢纽的各种建筑物与设备,控制并调节河川天然径流,达到除水害、兴水利的目的,亦称水库的控制运用。

 Writing

 通知(Notice)

通知是上级对下级、组织对成员部署工作、传达事情或召开会议时所使用的一种文体。通知有两种:一种是以布告的形式把事情通知有关人员,如员工、教师、会员、读者、观众等;另一种是以书信的形式把事情传达给有关人员。英文通知的格式与中文通知大体相同,由标题、正文、落款三部分组成。

标题写在正文上方正中位置。落款即发出通知的单位和时间,可写在正文右下角,有时这项也可省略。书写正文时应注意以下几点:内容应简洁清楚;多采用被动语态,人称多用第三人称。常见的通知有会议通知、讲座或报告通知、活动通知、招聘通知、征稿通知等。

会议通知

NOTICE

Head of all sections are requested to meet in the Meeting Hall on Friday at 2:00 p.m. to attend a meeting on the financial program of the company. Each participant should hand in a written proposal.

The Administrative Office

通知

定于星期五下午 2 点在会议厅召开公司财政会议，望各部门领导准时出席。每位与会者须提交一份书面建议。

参考用语（reference expression）

1. All are warmly welcome.
热烈欢迎大家参加。
2. We are pleased to inform that…
很高兴通知大家……
3. This is to inform\notify that…
特此通知。
4. Please be informed that…
通知
5. A lecture will be held in\at …(place) on …(date)…
将于……时间……地点举行会议。
6. All are expected\requested to attend the meeting.
希望\要求所有人参加会议。

专业英语中常用的词头

词头	意义	词例
1. ab-	脱离	abnormal 变态的,反常的
2. ad-	加添,到	adjust 调整;adjoin 接,邻接
3. anti-	反对,抗	antisymmetric 非对称的;antirust 防锈的
4. auto-	自己,自	autotransformer 自耦变压器
5. bi-	双,二	biphase 双相;bisect 等分为二
6. by-	附属的,次要的,旁边的	by-product 副产品,副业
7. centi-	百,百分之一	centigrade 百分度的,厘米
8. circum-	环境,在周边	circumference 圆周,外切
9. co-	一起,共,和	co-exist 共存;cohesion 凝聚,黏聚力
10. con-,com-	共同,一起	concur 同时发生;combine 联合
col-,cor-	共同,一起	collect 收集;correspond 符合,对应
11. contra-	反对,相反	contradiction 矛盾;contrary 相反的
12. counter-	反,逆	counterclockwise 反时针的(地)
13. cross-	横,交叉,十字	cross-section 横断面;crossroad 交叉路

14. de-	除去	decompress 排除压力;decrease 减少	
15. deca-	十	decameter 十米;decade 十年	
16. deci-	十分之一,分	decimeter 分米;decimal 小数的,十进制的	
17. di-	二,双,偶	dioxide 二氧化物	
18. dia-	通过,横过	diameter 直径;diagonal 对角线	
19. dis-	不,无,解除	disconnect 切断,分开;displace 排水,移动	
20. electro-	电,电气	electrodynamics 电动力学	
21. en-	置于,使	envelope 信封,封皮;enclose 封入;enable 使能	
22. equi-	同等	equivalent 等量,等值,等量的,等值的	
23. ex-	除去,离开,出自	exclude 除掉,除……外	
24. fore-	前,先,预	forecast (大气)预报;foreword 绪言,前言	
25. ferro-	铁,钢	ferro-concrete 钢筋混凝土	
26. hecto-	百	hectometer 百米	
27. hex(a)-	六	hexagon 六角形	
28. homo-	同	homogeneous 均匀的;homophase 同相	
29. hydro-	水,氢	hydro-power 水力;hydrostatics 流体静力学	
30. hyper-	超越	hypersonic 超声速的;hyperpressure 超压	
31. hypo-	低,次	hypofunction 机能减退	
32. im-, in- il-, ir-	不,非 不,无非	impure 不纯的;inelastic 非弹性的 illogical 不合逻辑的;irregular 不规则的	
33. in-	在内,向内	inlet 入口;input 输入(量)	
34. inter-	相互,间,中间	interaction 相互作用;intercross 交叉	
35. intro-	向内,向中	introduce 引导,传入	
36. iso-	等,同	isobar 等压线;isogonal 等角的	
37. kilo-	千	kilometer 千米	
38. macro-	大,宏(观),常量	macroscopic 宏观的,肉眼可看的	
39. meg(a)-	大,兆,百万	megacycle 兆周	
40. micro-	微,百万分之一	microwaves 微波;micrometer 千分尺	
41. mid-	中	middle 中间(的)	
42. milli-	千分之一,毫	milligram 毫克	
43. mis-	误	misunderstand 误差,误会;misuse 误用	
44. mon(o)-	单	monoxide 一氧化物	
45. multi-	多	multiphase 多相的;multinomial 多项的	
46. non-	非	non-metal 非金属,非弹性的	
47. nona-	九	nonagon 九角形	
48. oct-, octa-, octo-	八	octagon 八角形	

49. out-	超过,向外	output 产量,输出;outlet 排泄口	
50. over-	超过,过分	overload 超载;overflow 溢出	
51. pent(a)-	五	pentagon 五角形	
52. peri-	周围,近,环境	perimeter 周长;period 周期	
53. phono-	声,音	phonometer 测声计	
54. photo-	光	photo-cell 光电管	
55. poly-	多,聚,复	polygon 多边形;polymer 聚合物	
56. post-	在后,补充	postaxial 轴后的	
57. pre-	在前,预先	preceding 上述的;preset 预置	
58. pro-	向前	proceed 前进,进行;produce 生产,产生	
59. quadt-	四	quadrangle 四边形	
60. quinque-	五	quinquangular 五角形的	
61. re-	再次重复,返回	reconstruction 重建,反对;reaction 反作用,反应	
62. radio-	放射,辐射,无线电	radio-active 放射性的	
63. self-	自动,因有,自	self-induction 自感应;self-enery 固有能量	
64. semi-	半	semi-conductor 半导体	
65. sex-	六	sexangle 六角形	
66. sept-	七	septangle 七角形	
67. sub-	次于,在下,低,再	submerge 浸在水中;subsonic 亚声速的	
68. super-	超	supersaturate 过饱和	
69. sur-	超,在上	surface 表面;surpass 超过	
70. syn-,sym-	同,共	synthesis 合成;symmetry 对称	
71. tele-	远	telescope 望远镜;telephone 电话	
72. tetr(a)-	四	tetragon 四角形	
73. thermo-	热	thermodynamic 热力学	
74. trans-	横过,转移	transport 运输;transformer 变压器	
75. tri-	三	triangle 三角形	
76. ultra-	超,极端	ultra-sonicwave 超声波	
77. un-	不	unable 不能的;unequal 不相等的	
78. uni-	单,一	unit 单位,单元;uniform 均匀的	
79. under-	不足,在……下	undercharge 充电不足;underwater 水下的	
80. with-	反对,返回	withdraw 撤回;withstand 抵抗,经得起	

Exercise 1: Match the words and expressions in the left column with the Chinese in the right.

1. hydrological phenomenon   A. 防洪调节
2. hydroelectric engineering   B. 径流调节
3. dynamic resource   C. 水利工程
4. water circulation   D. 年径流
5. hydrological stations   E. 水文统计
6. hydrological statistics   F. 水文现象
7. annual runoff   G. 年输沙量
8. runoff regulation   H. 水文站
9. annual sediment   I. 水循环
10. flood control regulation   J. 动态资源

Exercise 2: According to the content of Text 1, fill in the blanks.

Water cycle includes rivers, _____, _____, _____, runoff, and other links.

Hydrological information collected includes precipitation, surface evaporation, _____, _____, sediment, and _____, low water and storm survey.

Statistical law of hydrological phenomena is sought through mathematical statistics called hydrological statistics in hydrology, which includes _____ and _____.

Design flood is the basis of designing of hydraulic structures, which consists of three elements: the design _____, total design flood and design _____.

_____ includes profiting regulation and flood control regulation.

UNIT TWO Hydraulic structures
水利水电工程建筑物

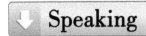

💧 **Weather and environment** 天气和环境

1. It is going to snow (hail) tomorrow, some measures must be taken to prevent freezing.

明天将下雪(冰雹),为预防冰冻必须采取一些措施。

2. The lifting work on site will be compelled to stop, owing to the strong wind (a dense fog).

由于强风(浓雾),现场起重吊装工作将被迫停止。

3. The weatherman says, the highest temperature during the day will be twenty one degrees centigrade (21 ℃).

天气预报员说:今天白天最高气温为 21 ℃。

4. The temperature will drop to five above (below) zero tonight.

今晚温度将降到零上(下)5 ℃。

5. There is an equipment in front of the building (behind the water tower, under the pipe rack, on the floor, inside the steel structure, in the workshop).

在建筑物前面(在水塔后面、在管廊下面、在地面上、在钢结构里面、在车间内)有一台设备。

💧 **Introduction of Itaipu Dam**

A (Foreign engineer)   B (Chinese engineer)

A: Hello!

B: Hello, what are you doing?

A: I'm looking at the images of Three Gorges Dam.

B: Ah, it's really a great engineering project.

A: Yes, do you know Itaipu Dam?

B: A little. It seems to be a very large hydropower station.

A: Tell you what, it's the largest operating hydroelectric facility in terms of energy generation.

B: Really?

A: Of course. Last year it generated 98.6 TWh, while the annual energy generation of Three Gorges was 83.7 TWh.

B: What a surprise! I thought the Three Gorges Dam was the largest power station.

A: Yes, you are fairly right. But that's about the installed capacity. In terms of energy generation, Itaipu Dam is the first.

B: Oh, I get it.

A: It is a binational undertaking run by Brazil and Paraguay at the Paraná River on the border section between the two countries.

B: Well, great! It's really one of the great engineering achievements.

Text 2 Hydraulic structures

Hydraulic engineering is built to control and allocate the natural surface water and groundwater for the purpose of promoting benefits and removing evils. For the purpose of promoting benefits, hydraulic engineering is mainly divided into flood control works, irrigation and water conservancy project, hydropower engineering, power supply and drainage engineering, navigation engineering, etc. For the purpose of comprehensive utilization of water resources to meet the need of flood control, irrigation, power generation, water supply and navigation, it is necessary to build various of structures to form an interrelated whole to control and distribute water together, constituting a comprehensive body called hydro junction, the buildings in which are known as hydraulic structures.

2.1 The Gravity dam on the rock foundation

Gravity dam keeps stable depending on the sliding resistance force generated by its self weight. Gravity dam is usually built on the rock foundation and made of concrete or masonry. The dam axis is generally straight. Permanent transverse joints are provided perpendicular to the dam, by which the dam is divided into several independent blocks to adapt to changes in temperature and inhomogeneous settlement of foundation.

2.2 The embankment and dike

Embankment is a kind of water retaining dam, composed by soil materials, stone materials or mixtures of them, heaped up by the methods of dumping fill and compaction, etc. Embankment keeps its self stability by the friction between the particles, the cohesive properties and compactness to resist the water pressure and prevent seepage damage.

2.3 Other dam types

Arch dam is a spatial shell structure. The dam body structure can be approximated seen as a series of horizontal arch rings and vertical cantilever beams. The dam structure has the roles of both the beam and the arch with the characteristics of two-way transferring of load. The rubber dam is a new type of hydraulic structure with the development of polymer composite materials industry. Under the combined action of technical heavy vibration and compaction, the compacted concrete dam is completed by vibrating and compacting on a thin layer paving, dry concrete mixes of relatively low cement content. The masonry dam has been widely used in middle and small sized hydraulic projects in the stone-rich areas.

2.4 Sluice

Sluice is a kind of low-head structure to control the flow and regulate the discharge, which has a dual role in retaining water and releasing water. Sluices are mostly built on the soil foundation.

2.5 The riverside spillway

The spillway is a kind of the most common release structures, which is built to release the excessive reservoir water and empty reservoirs for construction diversion when necessary, in order to meet the requirements of safety or others.

2.6 The water conveyance structures

The water conveyance structures are a kind of buildings to transport water from the upstream to the downstream, for the purposes of irrigation, power generation and water supply, which can be divided into pressurized water conveyance (water diversion) buildings and non-pressure water conveyance (water diversion) buildings.

2.7 Hydraulic transition process of hydropower station

Diversion system, water turbine and its speed control device, generators, electrical load of hydropower station compose a large power system. The system has two stable states: static and constant speed operation. When the power system is transferred from one state to another or is disturbed in a constant speed operation, the system will appear an unsteady transition (transient state) process, which will result in a series of engineering problems such as water hammer of pressure water pipe (channel), water level fluctuation of the surge chamber, speed change of water turbine set and the stability of the speed control system, etc.

2.8 Hydropower house

Hydropower station structures consist of intake structure, water conveyance structure, equalizing structure and powerhouse structure. Hydropower house is a comprehensive engineering facility transforming hydropower into electricity, which includes powerhouse buildings, water turbines, generators, transformers, switching stations and so on. It's also a place of production and activity for the operating personnel.

2.9 Layout of a hydro junction

Layout of a hydro junction is a relative concentration arrangement of various buildings with different roles to ensure a good corporation in operation.

主题2 水利水电工程建筑物

水利工程是指本着除害兴利的目的兴建的对自然界地表水和地下水进行控制与调配的工程。从兴利目的来看，水利工程主要分为防洪工程、农田水利工程、水力发电工程、供电和排水工程、航运工程等。为了综合利用水资源，达到满足防洪、灌溉、发电、供水和航运等目的，需要组合兴建多种不同类型的建筑物，形成一个相互联系的整体，共同控制和分配水流，由此构成的综合体称为水利枢纽，其组成的建筑物称为水工建筑物。

2.1 岩基上的重力坝

重力坝依靠坝体自重产生的抗滑力来维持稳定,通常修建在岩基上,用混凝土或浆砌石筑成。坝轴线一般为直线,垂直坝轴线方向设有永久性横缝,将坝体分为若干个独立坝段,以适应温度变化和地基不均匀沉陷。

2.2 土石坝与堤防

土石坝是指由土料、石料或土石混合料,采用抛填、碾压等方法堆筑成的挡水坝。土石坝主要利用土石颗粒之间的摩擦、黏聚特性和密实性来维持自身的稳定,抵御水压力和防止渗透破坏。

2.3 其他坝型

拱坝为一空间壳体结构,其坝体结构可近似看作由一系列水平拱圈和竖向悬臂梁组成,坝体结构既有梁的作用,又有拱的作用,具有双向传递荷载的特点。橡胶坝是随着高分子合成材料工业的发展而出现的一种新型水工建筑物。碾压混凝土坝利用机械的强力振动和碾压的共同作用,对分薄层摊铺的、水泥含量比较低的干硬性混凝土拌和料进行振动压实。浆砌石坝在石料丰富地区的中小型水利工程中得到广泛应用。

2.4 水闸

水闸是一种控制水流和调节流量的低水头建筑物,具有挡水和泄水双重作用,多修建在土质地基上。

2.5 河岸溢洪道

溢洪道是最常见的泄水建筑物,是为排泄水库多余水量、必要时放空水库以及施工期导流,以满足安全和其他要求而修建的建筑物。

2.6 输水建筑物

输水建筑物是为灌溉、发电和供水需要从上游向下游输水用的建筑物,分为有压输水(引水)建筑物和无压输水(引水)建筑物。

2.7 水电站水力过渡过程

水电站的引水系统、水轮机及其调速设备、发电机、电力负荷等组成一个大的动力系统,这个系统有两个稳定状态:静止和恒速运行。若动力系统从一个状态转移到另一个状态,或在恒速运行时受到扰动,系统都会出现非恒定的过渡(暂态)过程,由此产生一系列工程问题:压力水管(道)的水锤现象、调压室水位波动现象、机组转速变化和调速系统的稳定等问题。

2.8 水电站厂房

水电站组成建筑物为进水建筑物、输水建筑物、平压建筑物和厂房建筑物。水电站厂房是将水能转化为电能的综合工程设施,包括厂房建筑、水轮机、发电机、变压器、开关站等,也是运行人员进行生产和活动的场所。

2.9 水利水电枢纽布置

水利水电枢纽布置是将作用不同的建筑物相对集中布置,并保证它们在运行中良好配合地工作。

 Writing

 请柬(Invitation card)

请柬又称请帖或邀请书,一般分为正式和非正式两种。正式请柬的书写比较严谨,有一定的书面格式,主要用于举行婚礼、丧礼、宴会、舞会、招待会、展览会及重大会议的开幕式和闭幕式等。非正式请柬就是一般所说的邀请信或邀请条,主要用于亲戚朋友之间。

正式请柬没有称呼语和结束语,因为已经包含在正文中。收到请柬后,通常应立即回复,表示拒绝或接受。如果接受,则先表示谢意,然后写明出席时间和地点;如果拒绝,则先表明歉意,然后说明不能出席的原因。回柬在格式上应与请柬一致。

正式请柬一般包含以下内容:宾主姓名、客套话、活动时间、内容、地点及说明或附注。正式请柬的邀请者一般用第三人称,被邀请者也多用第三人称,间或也有用第二人称的。按照英美国家的习俗,邀请者如夫妇齐全且又住在一起,则把两者名字都写上。

如果邀请者希望被邀请者能及时答复或对被邀请者应邀时服装有所要求,可在请柬左下角或右下角加以注明。

婚礼请柬(Invitation to a wedding)

范例:
INVITATION CARD

Mr. and Mrs. John Smith Request the honor of the presence of

Mr. and Mrs. Brown

   at the marriage of their daughter

Elizabeth Smith

to

Mr. John Frederick Hamilton

   Saturday, the twenty-ninth of September

单元二 水利水电工程建筑物 17

　　at four o'clock p. m.
　at Church of Heavenly Rest
　　New York
布朗先生和夫人:
兹定于9月29号(星期六)下午4点在纽约天安教堂为小女伊丽莎白·史密斯与约翰·弗雷德里克·汉密尔顿先生举行婚礼,届时恭请光临。
约翰·史密斯夫妇鞠躬

参考用语(Reference expression)

1. Request the honor of the presence of…
恭候光临
2. Request the pleasure of your company at a small dance
敬请光临小型舞会
3. Please reply to …
请予赐复
4. On the occasion of the tenth anniversary of…
为庆祝……成立十周年

 备忘录(Memo)

　　备忘录主要用于提醒、督促对方就某一问题提出意见,是一种记录诸如情况汇报、问题处理、责任分工等内容以备忘的文体,多用于公司、机构等内部书面往来活动。备忘录可通过书面形式发送,也可通过填写单位统一印制的表格来完成。
　　备忘录一般由以下两大部分组成:
　　1. 开头 Heading
　　　To:收件人
　　　From:发件人
　　　Date:日期
　　　Subject:主题
　　2. 内容 Messages
　　例:
　　To:Mr. Lee ,leader of the supply section
　　From:Li Hua
　　Date:December 20,2005
　　Subject:Borrow a DVD player
　　Dear Sir ,
　　I'm the secretary of the Marketing Office and we are badly in need of a DVD player in order to have the programs for Christmas Party rehearsed. I hope you will

issue one to us as soon as possible.
　　With kind regards.
　　Yours sincerely,
　　Li Hua

致：李先生，供应处处长
自：李华
日期：2005年12月20日
事由：借一部DVD机
亲爱的先生：
　　我是营销办公室秘书，我们急切需要一部DVD机以排练圣诞晚会节目。希望你能立即发给我们一部。谨致
　　问候
　　您诚挚的
　　李华

 词义的处理

一、词义选择

在英语词组中词的搭配方面，定冠词和介词常起着重要的作用，使整个词组意义改变，因此在翻译英语时，切忌望文生义，按汉语思考，想当然地随意进行翻译，例如下列这些词组：

in course of　　在……过程中
in the course of　　在……期间
in case of　　如果，在……时；在……情况下
in the case of　　就……而论；在……情况下
in the case of　　如果
in consideration of　　在考虑……时
in the consideration of　　面临
in face of　　当……之前；不顾
in the front of　　在……前部
out of question　　毫无疑问；无疑地
out of the question　　不可能的
think over　　仔细考虑
think about　　考虑

二、词义引伸

在科技英语中词义常以引伸或被赋予专门意义的一些词的例子。

例词　常用词义；可能引伸的词义

activity　活动性,组织；机构
backup　支持；阻塞,备用设备
creep　爬行,徐变；渗水
cutoff　切去,截弯取直；截水
development　发展,研究；开发工程；成就
factor　因素；问题
flexibility　柔性；易弯性,适应性
hardware　金属构件；设备,硬件
software　设计计算方法；方案,软件
joint　连接；接缝
loading　装载,荷载；负荷
output　产量,出力；输出
philosophy　哲学；自然科学
project　规划；方案,工程；课题项目
snowballing　滚雪球式的扩大；迅速增长
work　工作,功；加工
yielding　屈服；塑性变形
exceeded　超过；不能被满足
historically　在历史上；从历史发展看
increasingly　愈加；日益,日趋完善
literally　照字义；逐字地,可毫不夸张地说
proportional　成比例的；尺寸相称的
appreciably　可估计地；明显地

Exercise 1: Match the words and expressions in the left column with the Chinese in the right.

1. hydro junction　　　　　A. 重力坝
2. hydraulic structures　　　B. 土石坝
3. embankment　　　　　　C. 拱坝
4. the rubber dam　　　　　D. 浆砌石坝
5. the masonry dam　　　　E. 水利枢纽
6. the gravity dam　　　　　F. 水工建筑物
7. arch dam　　　　　　　G. 溢洪道
8. sluice　　　　　　　　　H. 橡胶坝

9. spillway   I. 输水建筑物
10. water conveyance structures J. 水闸

Exercise 2: According to the content of Text 2, fill in the blanks.

1. _____ keeps stable depending on the sliding resistance force generated by its self weight. It is usually built on the rock foundation and made of concrete or masonry.

2. _____ is a kind of water retaining dam, composed by soil materials, stone materials or mixtures of them, heaped up by the methods of dumping fill and compaction, etc.

3. _____ is a spatial shell structure. The dam body structure can be approximated seen as a series of horizontal arch rings and vertical cantilever beams.

4. _____ is a new type of hydraulic structure with the development of polymer composite materials industry.

5. _____ has been widely used in middle and small sized hydraulic projects in the stone-rich areas.

UNIT THREE Irrigation and drainage engineering
灌溉排水工程技术

Speaking

 Technical documents and drawings 技术资料和图纸

1. We completed this task according to the drawing number SD-76.

我们按照图号 SD-76 的图纸完成了这项工作。

2. According to the technical standard (norm, rules of operation), the erection (alignment, testing) work is now getting on.

安装(校准、试验)工作正在根据技术标准(规范、操作规程)进行。

3. This is a plot plan (general layout, general arrangement, detail, section, erection, flow sheet, PID, assembly, civil, electrical, control and instrumentation, projection, piping, isometric) drawing.

这是一张平面布置(总平面、总布置、细部、剖面、安装、流程、带仪表控制点的管道、装配、土建、电气、自控和仪表、投影、配管、空视)图。

4. That is a general (front, rear, side, left, right, top, vertical, bottom, elevation, auxiliary, cut-away, birds eye) view.

那是一张全视(前视、后视、侧视、左视、右视、顶视、俯视、底视、立视、辅助、内部剖视、鸟瞰)图。

5. How many drawings are there in the set?

这套图纸有几张?

6. Is this a copy for reproduction?

这是一份底图吗?

7. What is the edition of this drawing?

这张图纸是第几版?

8. Is this drawing in effect?

这张图纸有效吗?

9. Is this a revised edition?

这是修订版吗?

10. Will it to be revised yet?

这张图还要修订吗?

11. Are there some modifications(revisions) on the drawing?

这张图上有些修改(修正)吗?

12. The information to be placed in each title block of a drawing include: drawing number, drawing size, scale, weight, sheet number and number of sheets, drawing title and signatures of persons preparing, checking and approving the drawing.

每张图纸的图标栏内容包括:图号、图纸尺寸、比例、重量、张号和张数、图标,以及图纸的制图、校对、批准人的签字。

13. There are various types of lines on the drawing such as: border lines, visible lines, invisible lines, break lines, phantom lines.

图上有各种形式的线条,诸如:边框线、实线、虚线、断裂线、假想线。

14. We have not received this drawing(instruction book, operation manual), please help us to get it.

我们还未收到这张图纸(说明书、操作手册),请帮助我们取得。

15. Please send us further information about this item.

请将有关这个项目的进一步资料送交我们。

16. I want additional information on this.

我需要这方面的补充资料。

17. Please explain the meaning of this abbreviation(mark, symbol) on the drawing.

请解释图上这个缩写(标记、符号)的意义。

18. We comply with and carry out the GB(ANSI, BS, AFNOR, JIS, and DIN) standard in this project.

在这个工程中我们遵守并执行中国国家标准(美国标准、英国标准、法国标准、日本标准、联邦德国标准)。

19. Please make a sketch of this part on the paper.

请将这个零件的草图画在纸上。

20. This is a translated information, it maybe not quite sure.

这是翻译的资料,可能不太确切。

21. It is a mistake in translation.

那是一个翻译错误。

22. Please give us a copy of this information (technical specification, instruction, manual, document, diagram, catalog).

请给我们一份这个资料(技术规程、说明书、手册、文件、图表、目录样本)的复印本。

23. Please send us a technical liaison letter about it.

请给我们一份有关此事的技术联络单。

24. A working drawing must be clear and complete.

工作图必须简明、完整。

25. Data on equipment can be found in the related information.

设备的数据可从有关的资料中找到。

26. Have you any idea how to use the manufacturer's handbook?

你知道怎样使用这本厂家手册吗?

Construction machinery

A(Chief of section)   B(Director)

A: We have all kinds of construction machinery on the job site.

B: Yes, I see. So many! Are they all necessary?

A: Certainly. All of them are in quite different roles.

B: Oh, I know, the excavator is such an important machine. The invention of it greatly improves the constructing speed compared with the days before.

A: Exactly! The use of it creates a new age for the whole world.

B: Then what is the horsepower(HP) of this engine?

A: 148 horsepower. Look, this is the concrete mixer made by Hua-Dong Works.

B: Can you tell me the operation of this machine?

A: Sorry, I don't know, either. It's usually operated by our specialists. I just know it is steady in quality and reliable in performance.

B: Glad to hear that! A good machine.

A: But to maintain its efficiency, we have to ensure a regular service.

B: That's quite reasonable. Generally speaking, any machine needs a service.

A: You said it. To tell the truth, we have so much work to do to maintain all these machinery every day.

B: I think so.

A: Oh, it's time for lunch.

B: OK, Let's go.

Text 3　Irrigation and drainage engineering

Irrigation and drainage technology is a comprehensive science and technology, which is aiming at regulating the farmland water status, improving regional hydrological variation, using effective regulation measures scientifically and reasonably to eliminate flood and drought disasters, and to service agricultural production and sound development of the ecological environment by utilizing water resources rationally.

The service object mainly includes two aspects, one is regulating farmland moisture conditions and the second is improving and regulating the regional hydrological condition.

The basic contents of irrigation and drainage engineering techniques include as follows: analysis and determination of water requirement rule and water demand, determination of irrigation process and water consumption, irrigation methods and irrigation techniques, rational utilization of water resources in the agriculture field, water supply method, project layout and design of water channel (or pipeline).

3.1　Farmland moisture conditions and irrigation water consumption in the farmland

There are three kinds of existing forms of farmland moisture, i.e. surface water, soil water and groundwater, among which soil water has the closest relation with the crop growth. Farmland moisture condition is mainly divided into two kinds, i.e. dry land farming region and rice region. Irrigation water use and irrigation water discharge are the water quantity and water flow introduced from the water source. They are the indispensable data of watershed planning and regional water planning.

3.2 Surface irrigation technology

Irrigation method is defined as the pattern of the irrigation water entering the field and moistening the plant root zone soil. Irrigation technology refers to a series of scientific measures corresponding to an irrigation method. Traditional surface irrigation technology includes border, furrow and basin irrigation. Ground water-saving irrigation technology includes surge irrigation, small furrow "three changes" irrigation, long furrow sectional irrigation, and combined the wide and shallow furrows irrigation.

3.3 Irrigation water source and water intake engineering technology

Irrigation water source is the natural resource that can be used for irrigation. The main types are surface irrigation water source, underground irrigation water source, rainwater harvesting, the development and utilization of poor quality water, and combined use of different water sources. There are four kinds of water intake approaches, i. e. diversion without dam, diversion dam, water intake by pumping and water intake from reservoir.

3.4 Planning and design of irrigation channels

The irrigation canal system is composed of irrigation channels at all levels and the outlet (discharge) channels. The layout should meet the requirements of the overall design and irrigation standards. The structure of a canal system is a building which is constructed on the channel system in order to meet the needs of various departments in a safe and reasonable way. According to the function can be divided into control buildings, crossing buildings, drainage structures, connecting buildings, water measuring buildings, etc. .

3.5 Channel seepage control

Channel seepage control engineering technologies refer to various engineering technical measures to reduce channel leakage loss. The channel impermeable materials mainly include soil, cement soils, stone, coating material, concrete, bituminous concrete, etc. Anti-frost heave measures in the channel seepage control engineering involve avoiding frost heave method, reducing frost heave method and optimizing structure method.

3.6 Well irrigation district planning

On the basis of the comprehensive analysis and summations of various basic data, well irrigation district planning is put forward according to the planning principles and tasks. The well irrigation district planning can be divided into as follows in accordance with main tasks: ①the new well irrigation district planned to develop; ②reconstruction and planning of the old well irrigation district; ③the well irrigation district combined well and channel; ④the well irrigation district needing comprehensive management of preventing water logging and solotization.

3.7 Sprinkling irrigation engineering technology

Sprinkling irrigation is a irrigation method by taking use of sprinklers and other special equipments to spray pressurized water into the air, and forming water droplets and falling to the ground and crop surface. Sprinkling irrigation system mainly consists of water source engineering, water pump and power equipment, transmission and distribution pipe network systems, sprinklers and auxiliary works, auxiliary equipment and other components.

3.8 Micro irrigation engineering technology

Micro irrigation is an irrigation method of transforming pressurized water into small stream or droplet by special equipment and moistening the soil near the crop root, including drip irrigation, microspray irrigation and spring irrigation, etc. Micro irrigation system is composed of water source works, head works, piping system and douche.

3.9 Low pressure pipe irrigation engineering technology

Low pressure pipe irrigation project is a form through superseding open channel irrigation by pipe. Under a certain pressure, the irrigation water is transported to the field by outlet facility, then to the ditch and furrow by the pipe outlet or exterior water pipe. By comparing with other irrigation methods, it has several advantages: ①saving water and energy; ②saving area and labor; ③low cost and high efficiency; ④high adaptability and management convenience. Low pressure pipe irrigation system consists of water source and intake works, transportation and distribution pipe network system and field irrigation system.

3.10 Field drainage system

Farmland drainage is to collect and exclude excessive water, reduce and control the water table, improve crop growth environment, prevent and eliminate the disasters of flood, water logging, and salinization, provide a good environmental conditions for the normal growth of crops. According to the spatial position, the field drainage system can be classified into two kinds, horizontal drainage and shaft drainage. According to the patterns of field drainage, the field drainage includes ditch drainage systems, buried pipe drainage system and shaft drainage system.

3.11 The plan and design of core drainage system

Drainage channel and pipe system has the characteristics of wide distribution, large quantity, and important influence. The criteria of planning layout should be followed are lower layout, reasonable economy, row distribution by elevation, overall planning and comprehensive utilization. Drainage system is generally divided into two basic types: ① general drainage system; ② comprehensive utilization drainage system.

主题3 灌溉排水工程技术

灌溉排水技术是调节农田水分状况和改善地区水情变化,科学合理地运用有效的调节措施,消除水旱灾害,合理利用水资源,服务于农业生产和生态环境良性发展的一门综合性科学技术。

灌溉排水技术的服务对象主要包括两个方面:一是调节农田水分状况,二是改善和调节地域水情。

灌溉排水工程技术的基本内容包括:分析和确定作物的需水规律和需水量,灌溉用水

过程和用水量的确定，灌溉方法和灌水技术，水资源在农业方面的合理利用，水源的取水方式，输水渠道（或管道）工程的规划布置及设计。

3.1　农田水分状况和农田灌溉用水量

农田水分存在地面水、土壤水和地下水三种形式，其中土壤水是与作物生长关系最密切的水分存在方式。农田水分状况主要分为旱作地区和水稻地区两种。灌溉用水量和灌溉用水流量是指灌区需要从水源引入的水量和流量。它们是流域规划和区域水利规划不可缺少的数据。

3.2　地面灌水技术

灌水方法是指灌溉水进入田间并湿润植物根区土壤的方式与方法，而灌水技术则是指相应于某种灌水方法所采取的一系列科学措施。传统地面灌水技术有畦灌、沟灌和淹灌。节水型地面灌水技术有波涌灌、小畦"三改"灌水技术、长畦分段灌和宽浅式畦沟结合灌水技术。

畦灌

沟灌

淹灌

波涌灌

3.3　灌溉水源及取引水工程技术

灌溉水源是指天然资源中可用于灌溉的水体，主要类型有地表灌溉水源、地下灌溉水源、雨水集蓄、劣质水开发利用和不同水源的联合利用。灌溉取水方式有无坝引水、有坝引水、抽水取水和水库取水。

3.4　灌溉渠道系统规划设计

灌溉渠道系统由各级灌溉渠道和退（泄）水渠道组成。其布置应符合灌区总体设计和灌溉标准要求。渠系建筑物是指为安全、合理地输配水量，以满足各部门的需要，在渠道系统上所建的建筑物，按作用可分为控制建筑物、交叉建筑物、泄水建筑物、衔接建筑物、量水建筑物等。

3.5 渠道防渗

渠道防渗工程技术是指为减少渠道渗漏损失而采取的各种工程技术措施。渠道防渗按材料分为土料、水泥土、石料、膜料、混凝土、沥青混凝土等。渠道防渗工程的防冻胀措施有回避冻胀法、削减冻胀法和优化结构法。

3.6 井灌区规划

井灌区规划是在综合分析与归纳区内各种基本资料的基础上，根据规划原则，结合规划任务的需要所得出来的结果。井灌区规划按其主要任务不同可分为：①计划发展的新井灌区；②对旧井灌区的改建规划；③井渠结合的井灌区；④防渍涝和治碱等综合治理的井灌区。

3.7 喷灌工程技术

喷灌是一种利用喷头等专用设备把有压水喷洒到空中，形成水滴落到地面和作物表面的灌水方法。喷灌系统主要由水源工程、水泵及动力设备、输配水管网系统、喷头和附属工程、附属设备等部分组成。

3.8 微灌工程技术

微灌是一种转化成小液滴流或加压水通过特殊设备和湿润作物根系附近土壤的灌水方法，包括滴灌、微喷灌和微灌灌溉等。微灌系统由水源工程、喷头、管道系统和冲洗组成。

3.9 低压管道输水灌溉工程技术

低压管道输水灌溉工程是以管道代替明渠输水灌溉的一种工程形式。通过一定的压力，将灌溉水由分水设施输送到田间，再由管道分水口或外接管输水进入田间沟、畦。它与其他灌溉方式比较有几个优点：①节水节能；②省地、省工；③成本低、效益高；④适应性强、管理方便。低压管道输水灌溉系统由水源与取水工程、输水配水管网系统和田间灌水系统组成。

3.10 田间排水系统

农田排水就是汇集和排除农田中多余水量,降低和控制地下水位,从而改善作物的生长环境,防治和消除涝、渍及盐碱灾害,为作物的正常生长创造良好的环境条件。田间排水系统按空间位置可分为水平排水和竖井排水两大类。根据田间排水方式的不同,田间排水系统有明沟排水系统、暗管排水系统和竖井排水系统三种方式。

3.11 骨干排水系统规划设计

排水沟道系统分布广、数量多、影响大,在规划布置时应遵循低处布置、经济合理、高低分排、统筹规划和综合利用几个原则。排水系统一般分为两个基本类型:①一般排水系统;②综合利用的排水系统。

Writing

求职信

求职信属于正式信函,一般分为三段:

第一段说明得知信息的渠道并说明写信的目的;

第二段介绍本人的年龄、学历、才能与胜任此职有关的历史和资格,要突出履历中的精华;

第三段希望聘方给予考虑并提供联系方式。

写求职信的常用套话如下:

1. I wish to apply for the position of cashier advertised in today's China Daily.

今天《中国日报》有招聘出纳员的广告,我想申请这一职位。

2. I am very interested in the position, and I'm trying to apply for it.

对此职位很有兴趣,我想申请这一职位。

3. In reply to your advertisement in today's…(newpaper) for a…, I hasten to write this letter of application for the post.

看到你们在今天……报纸上征聘一名……的启事,我赶紧写这封信以谋求这一职位。

4. I believe that my ability and experience will fully qualify me for the position of…

我相信我有能力与经验担任……这份工作。

5. I am enclosing my CV/personal history which I believe will show you why I feel I can meet the requirements of a cashier in your company.

现随函附上简历,我相信它会证明我能满足贵公司对出纳员的各种要求。

6. Have you an opening in your company for a young man who believes that he has something to offer?

有位年轻人认为他能够对贵公司有所贡献,贵公司能对他打开大门吗?

7. I should be grateful/much obliged if you could give me an opportunity to try to serve you in the position.

如果能给我机会在这一岗位上为您效劳,我将不胜感激。

8. I hope you will take my case into your consideration.

希望您能考虑我的申请。

参考范文

Dear Sirs:

I have just seen your advertisement in Beijing Evening News of the 6th August for a salesman in the Electronic Device Section of your company. I'm very interested in the job and I feel I'm qualified to meet the requirements. I'm therefore enclosing a resume together with reference from my supervisor.

As you can see, I once worked in the Electronic Department. So I am familiar with different kinds of electric devices. I have worked more than once as a salesman in some stores during my previous vacations. Besides, I'm very patient and friendly in nature. I'm confident that I shall be suitable for the kind of job.

If you need any further information, I shall be very pleased to supply it. Or I wonder if you will grant me with an interview.

I'm looking forward to hearing from you soon.

Yours faithfully
Wu Qi

科技术语翻译

一、意译法

意译法即是根据英语原词的意义译成适当的汉语。在意译时,术语中词头、词尾的意义,对术语的定名有着相当重要的作用,例如:

televisor 电视机;semiconductor 半导体

electronic 电子学;microelectronics 微电子学

video-recorder 录像机;video-phone 电视电话

这些例词中的词头 tele-(远)、semi-(半)、electro-(电)、micro-(微);词尾-er, or (物),-ics(学术),-phone(声音)等是明显的例子。

二、音译法

在科技术语中,下列几种范畴采取音译法:

(1)某些计量单位。例如:

Ohm 欧姆;Calorie 卡路里;Ampere 安培;Volt 伏特;Joule 焦耳;Henry 亨利(电感单位);Newton 牛顿;Watt 瓦特;Pascal 帕斯卡

(2)某些科技发明、材料名称及人名。例如:

radar 雷达;laser 莱塞(激光);sonar 声纳;pump 泵(浦);engine 引擎(发动机); vitamin 维他命(维生素);nylon 尼龙;microphone 麦克风;aspirin 阿斯匹林,吗啡;vaseline 凡士林;penicillin 盘尼西灵(青霉素);Archimedes 阿基米德;Bernoulli 伯努利;Euler 欧拉

三、形译法

在科技英语术语中用英文字母表示某种事物的外形,翻译时应选用近似这种字母形状的汉语词汇来表达其外形,有时也保留原字母不译。

(1)选用近似这种字母形状的汉语翻译。例如:

I-bar 工字铁;I-section 工字形截面

T-beam 丁字梁;T-Socket 丁字套管

U-bolt 马蹄螺栓;U-steel 槽钢

V-belt 三角皮带;Y-pipe 叉形管

(2)保留原字母不译,以字母表达形状。例如:

C-clamp C 形夹;A-frame A 形架

U-nut U 形螺母;O-ring O 形环

T-joint T 形接头;Z-axis Z 轴

Y-connection Y 形连接;Z-crank Z 形曲柄

(3)保留原字母不译,以字母代表一种概念,例如:

AT-cut AT 切片;Q-antenna Q 天线;X-ray X 射线

四、意、音兼顾

ampere-meter 安培表;valve-guide 阀导,汽门导管
valvebody 阀体;engine case 引擎箱
Curie point 居里点;Einstein equation 爱因斯坦方程
Archimedes principle 阿基米德定律
Bernoulli equation 伯努利方程

Exercise 1: Match the words and expressions in the left column with the Chinese in the right.

1. irrigation and drainage technology   A. 地面灌水技术
2. farmland moisture conditions   B. 畦灌
3. surface irrigation technology   C. 波涌灌
4. border irrigation   D. 农田水分状况
5. surge irrigation   E. 灌溉排水技术
6. channel seepage control   F. 灌溉渠道系统
7. micro irrigation   G. 长畦分段灌
8. long furrow sectional irrigation   H. 渠道防渗
9. drainage channel and pipe system   I. 微灌
10. sprinkling irrigation   J. 喷灌

Exercise 2: According to the content of Text 3, fill in the blanks.

1. There are three kinds of existing forms of farmland moisture, i. e. _____ water, _____ water and _____, among which soil water has the closest relation with the crop growth.

2. Traditional surface irrigation technology includes _____, _____ and _____ irrigation.

3. There are four kinds of water intake approaches, i. e. _____, diversion dam, _____ and water intake from reservoir.

4. _____ irrigation is a irrigation method by taking use of sprinklers and other special equipments to spray pressurized water into the air, and forming water droplets and falling to the ground and crop surface.

5. _____ irrigation project is a form through superseding open channel irrigation by pipe.

Construction organization and management of hydraulic projects
水利水电工程施工组织与管理

Speaking

 Engineering project 工程项目

1. A project execution is usually divided into some elementary phases, such as: engineering, procurement and transportation, and field construction.

一个工程项目的实施通常可分为几个基本阶段:工程设计、采购和运输,以及现场施工。

2. The contract number of this project is CJC78-8.

这个项目的合同号是 CJC78 – 8。

3. The seller(vendor) is Toyo Engineering Corporation(TEC) of Japan.

卖方(卖主)是日本的东洋工程公司(简称 TEC)。

4. The buyer (customer, client) is China National Technical Import Corporation(CNTIC).

买方(主顾、顾客)是中国技术进口总公司(简称 CNTIC)。

5. China National Chemical Construction Corporation (CNCCC) contracts for domestic and overseas chemical projects.

中国化工建设总公司(简称 CNCCC)承包国内和海外的化工工程。

6. Are you the Seller's Representative on the job site?

你是卖方的现场代表吗?

7. I am the Buyer's General Representative(GR).

我是买方的总代表(简称 GR)。

8. It is an inquiry(commercial and technical proposal, approval, agreement,

protocol, annex, technical appendix) about this project.

这是这个项目的询价书(商务和技术报价书、批准书、协议、会议记录、附加条件、技术附件)。

9. There is much information in the technical proposal, which including: process flow, process description, capacity of the plant, performance of the product.

技术报价书中有很多资料,包括工艺流程、工艺说明、生产能力、产品特性等。

10. The project team normally consists of project engineer, design engineer, schedule engineer, and various specialists.

项目工作组通常包括项目工程师、设计工程师、计划工程师以及各类专家。

11. We can evaluate the results of field construction by four criteria, which are quality, time, cost and safety.

我们可以通过四个指标来评价现场施工的成绩,即质量、时间进度、费用和安全。

12. I am responsible for the technical (scheduling, inspection, quality control) work of this project (area).

我负责这个项目(区域)的技术(技术、检查、质量控制)工作。

13. Would you tell us the technical characteristic about this project?

你能告诉我们有关这个工程项目的技术特性吗?

14. Please give a description about this project.

请对这个工程项目作一个叙述说明。

15. Do you have any reference materials about this project?

你有关于这个工程项目的参考资料吗?

对话

A(Supervisor)　　　B(Foreman)

A: What's written on the boom side?

B: "Danger! Keep away from under boom!"

A: Good, we should keep safety in mind all the time.

B: Yes, I agree. "Safety first".

A: Please take me to the power house.

B: Ok, this way please.

A: In general, you did a fairly good job. But there are still some problems you need to pay attention to.

B: Thank you so much for your visit. Your advice and suggestion on our work will be most valuable.

A: There should be more safety warning boards inside and outside the house.

B: Yes, we've realized that and it will be improved as soon as possible.

A: Have you ever seen some worker die in the house?

B: No, never.

A: That's what used to happen.

B: Really?

A: Of course, especially the front-line workers. They lack the related safety knowledge.

B: Exactly.

A: So you need to pay more attention to safety management, especially in the power house.

B: Thank you. I'll keep that in mind.

Text 4　Construction organization and management of hydraulic projects

Construction organization and management is to research the implementation process of construction and installation works. Reasonable construction and organization methods are found according to the construction organization of different projects. Construction conditions, construction quality, safety technology, and civilized construction are solved through scientific management to ensure a smooth implementation of the project. Some criterion should be followed in process: fully implementing greater, faster, better and more economical results, adhering to the basic construction procedures, rationally organizing and constructing according to the principles of systems engineering, carrying out the scientific management, balancing the human and material resources from the actual project, ensuring a continuous rhythmic construction.

Before construction organization and management, it is necessary to propose the project, and follow the construction process: undertaking engineering tasks, signing a construction contract, preparing for construction, proposing report of commencement, organizing construction, completion acceptance, placing in operation, putting into effect step by step. The construction preparation work includes preparatory work for construction contents, investigation and collection of information on the construction, technical data preparation, goods and materials preparation, construction site preparation, commencement condition and commencement report preparation.

4.1 Construction of hydraulic projects

According to their own characteristics and composition contents, the basic construction projects of water conservancy and hydropower project are divided into two types, five parts, and three-stage projects. According to the internal composition of the project itself, it can be divided into individual work, unit work, partitioned work and subdivisional work. Specific work of the construction include river watershed planning, proposing of project suggestions, feasibility study, preliminary design, construction preparation, construction and implementation, production preparation, completion and acceptance, post evaluation, etc. This process is a basic requirement for the national water conservancy and hydropower project construction, which can reflect the entire process of basic construction work of water conservancy and hydropower project. During the process of construction, a temporary specific organization engaged in engineering should be formed to accomplish specific tasks. The functions of the organization, forms, and project management modes should be well understood to achieve the project objectives by effective organization and management of construction.

4.2 Network planning technique

Network planning technique, also known as network planning, is a way to carry out production organization and management. The basic principle of network planning technology is application of network graphics to represent the sequence and relationship of various works in a plan, including double codes network planning and single code network planning. The time parameters can be calculated through the network diagram. The key work and key path will be found out. The optimum solution will be sought by constantly improving network planning to achieve maximum economic effect with a minimum consumption. The main contents of network planning optimization include duration optimization, duration-resources optimization, duration-cost optimization.

4.3 Design of construction organization

Construction organization design is an important measure of implementing scientific and economic management for the whole construction process of project to be built, which can fully consider all the conditions of the proposed project, draft a reasonable construction scheme, determine the construction sequence, construction methods, labor organizations and technical and economic organization measures, and make overall arrangements reasonably of manpower, materials, machinery and engineering schedule to guarantee the quality and time of construction works for delivering. According to the compiling object, construction organization design can be divided into the total construction organization design, construction organization design of individual work and construction organization design of partitioned (subdivisional) work. The total construction organization design is made up of general construction schedule, general construction layout and technical supply.

4.4 Construction project management

Construction project management is a systematic management method system of effectively planning, organizing, coordinating and controlling of project in accordance with its inherent logic rule. For construction project management, the object is project item, the base is project manager responsibility system, the goal is realization of project objective, and the condition is formation of project items market. The main works of construction project management include establishing the organization of construction project management, compiling the planning of construction project management, controlling the target of construction project, optimizing the allocation of resource and dynamically managing on production factors of construction project, managing the contract and information of construction project, organizing and coordinating construction field management. In addition, a good ministry of construction project manager should be organized. Departments and staffs are provided in accordance with the actual project situation to realize the goal of construction project management effectively and complete the tasks of construction project management.

4.5 Contract management of construction project

Construction contract is one of the economic contracts between the owners and engineering consulting companies, design organizations, construction organizations or other corresponding units, as well as between these units, which is a written agreement in order to clarify the bilateral economic relationships such as responsibilities, rights, and interests, etc. during various activities for the completion of projects construction. The contract is a criterion of both sides in order to constrain, punish, and settle disputes. Contract management is the main content of project management, which is an effective way to reduce construction costs, improve economic efficiency, prevent financial risk and ensure project quality. The main contents of the contract management are files and document management, meeting management, payment control, default disposal and construction claims.

4.6 Cost management of construction project

The construction project cost is the sum of all manufacturing expenses occurred in construction, and is the product cost of an enterprise. The cost should be strictly controlled and managed. The cost of construction project can be reduced as much as possible by planning, organizing, controlling and coordinating of construction projects in order to maximize economic benefits. It is necessary to finish the following works: improving cost target prediction to determine the cost control objectives, determining the cost control principles around the cost target, and searching effective ways to achieve the goal of cost control. The methods of cost control include controlling the cost by working drawing estimate, controlling the human and material resources consumption by construction estimate, controlling the cost by using the method of cost analysis table, controlling the cost of divisional project by using the method of synchronous tracking of cost and progress, controlling the cost by disbursement schedule, establishing visa system of project cost audit, strengthening management to control the cost, adhering to site management standardization, establishing the resource consumption accounting to implement intermediate control of resource consumption, regularly carrying out "three synchronizations" inspection to prevent the project cost from abnormal gain or loss.

4.7 Safety management of construction project

The central issue of safety management of construction project is to protect the safety and health of human in productive activity, to ensure a smooth production. It grandly includes: safety regulations, safety technique, and industrial health.

The basic principles should be followed, such as dealing with Five Relations and Six Basic Principles exactly, Three Simultaneousness and Five Simultaneousness, Four Left off, Three Synchronizations and security identification. The existing unsafe factors include unsafe behavior from people and objects. It is necessary to clarify the roles, objectives and requirements of the safety management system, to ensure the safety of the production system. In order to determine the safety measures of construction project, it is necessary to be aware of the compiling requirements and main contents of technical measures of construction safety, as well as to learn security incident treatment, safety education and training.

4.8 Progress management of construction project

Progress management of project is the management of the progress extent in each stage and the final deadline during the project implementation. Its purpose is to ensure the achievement of the overall objectives under the premise of time constraints, which is one of the important measures to ensure timely completion of the project, reasonable arrangement of resources supply, and saving the project cost. Progress management of project includes drawing up and controlling a project schedule. The main measures to control the schedule are organization measures, technical measures, contract measures, economic measures and information management measures, etc.

4.9 Quality management of construction project

Quality management is to determine the quality policy, objectives and responsibilities and to realize all the activities of management functions through quality planning, quality control, quality assurance and quality improvement in quality system. Quality management aims at that the quality of construction projects must meet the design requirements and quality standards of the contract completely, meanwhile achieve the expectations about functions and useful value from the development organization. Its main contents include the basic work of quality management, quality systems design (planning), organization system and regulations of quality management, tools and methods of quality management, quality sampling inspection and control methods, quality cost, economic evaluation and calculation of quality management.

单元四 水利水电工程施工组织与管理

主题 4　水利水电工程施工组织与管理

施工组织与管理研究的是建筑安装工程的实施过程。不同工程根据施工组织找到合理的施工与组织方法,通过科学管理解决施工条件、施工质量、安全技术、文明施工等方面问题,确保工程项目顺利实施。过程中必须遵循:全面贯彻多快好省,坚持按基本建设程序办事,按系统工程原则合理组织施工,实行科学管理,从实际工程出发平衡人力物力,保证施工连续、有节奏地进行。

施工组织与管理前,需拟建工程项目,遵循施工程序:承接工程任务,签订施工承包合同,做好施工准备工作,提出开工报告,组织施工,竣工验收,交付使用,一步步实施。其中,施工准备工作包括进行施工内容准备工作、调查收集有关施工资料、技术资料准备、物资准备、施工现场准备、开工条件及开工报告准备。

4.1　水利水电工程建设

水利水电基本建设项目结合自身性质特点和组成内容划分为两大类型、五个部分、三级项目。按项目本身内部组成,划分为单项工程、单位工程、分部工程和分项工程。其建设程序的具体工作内容包括:进行流域河段规划,提出项目建议书,进行可行性研究,初步设计,施工准备,建设实施,生产准备,竣工验收,后评价等。此过程是国家对水利水电工程建设的基本要求,反映了水利水电工程基本建设工作的全过程。建设过程中,为完成特定任务需临时组建从事工程具体工作的组织,了解组织的职能、形式、工程管理方式,进行有效的施工组织与管理,以实现项目目标。

4.2　网络计划技术

网络计划技术,也称网络计划,是进行生产组织与管理的一种方法。网络计划技术的基本原理是:应用网络图形来表示一项计划中各项工作的开展顺序及其相互之间的关系,包括双代号网络计划和单代号网络计划。通过网络图进行时间参数计算,找出计划中的关键工作和关键路线,通过不断改进网络计划,寻求最优方案,以最小的消耗取得最大的经济效果。其中,网络计划优化主要内容包括工期优化、工期－资源优化、工期－费用优化。

4.3　施工组织设计

施工组织设计是对拟建工程全过程实施科学、经济管理的重要手段,可以全面考虑拟建工程的各种施工条件,拟订合理施工方案,确定施工顺序、施工方法、劳动组织和技术经济组织措施,合理统筹安排人力、材料、机械及工程进度计划,保证建设工程按质、按期交付使用。施工组织设计按编制的对象可分为施工组织总设计、单项工程施工组织设计和分部(分项)工程施工组织设计。其中,施工组织总设计主要包括施工总进度、施工总体布置和技术供应三部分。

4.4 施工项目管理

施工项目管理是按照其内在逻辑规律,对工程进行有效计划、组织、协调和控制的系统管理方法体系。施工项目管理以工程项目为对象、项目经理负责制为基础、实现项目目标为目的、构成工程项目要素的市场为条件。施工项目管理的主要工作有:建立施工项目管理组织,编制施工项目管理规划,进行施工项目的目标控制,对施工项目的生产要素进行优化资源配置和动态管理,施工项目的合同管理,施工项目的信息管理,组织协调施工现场管理。此外,还需组建好的施工项目经理部,根据工程实际情况设置部门和配置人员,有效地实现施工项目管理目标,完成施工项目管理任务。

4.5 施工项目合同管理

工程承包合同是经济合同的一种,是业主与工程咨询公司、设计单位、施工单位或其他有关单位,以及这些单位之间,为明确在完成项目建设的各种活动中双方责、权、利等经济关系而达成的书面协议。合同是双方行为的准则,具有制约、惩罚、解决纠纷的作用。合同管理是工程项目管理的主要内容,是降低工程造价、提高经济效益、预防经济风险、保证工程质量的有效途径。合同管理的主要内容有文件与档案管理、会议管理、支付控制、违约处置与施工索赔。

4.6 施工项目成本管理

施工项目成本是指在施工中发生的全部生产费用的总和,是企业的产品成本,需对其进行严格控制和管理。通过计划、组织、控制和协调等活动尽可能降低施工项目成本费用,获得最大限度的经济利益。施工项目成本管理需完成好以下几项工作:搞好成本目标预测,确定成本控制目标,围绕成本目标确定成本控制原则,查找有效途径实现成本控制目标。其中,成本控制的方法包括:以施工图预算控制成本支出,以施工预算控制人力资源和物资资源的消耗,应用成本分析表法控制项目成本,应用成本与进度同步跟踪的方法控制分部项目工程成本,以用款计划控制成本费用支出,建立项目成本审核签证制度,加强管理控制成本,坚持现场管理标准化,建立资源消耗台账实行资源消耗中间控制,定期开展"三同步"检查防止项目成本盈亏异常。

4.7 施工项目安全管理

施工项目安全管理的中心问题是保护生产活动中人的安全与健康,保证生产顺利进行。宏观的内容包括安全法规、安全技术、工业卫生。应遵循的基本原则有:正确处理"五种关系""六项基本原则""三同时"和"五同时""四不放过""三个同步",安全标识。其中,存在的不安全因素包括人的不安全行为和物的不安全行为。需明确安全管理体系的作用、目标和要求,保证生产体系的安全。确定施工项目安全技术措施,我们需要了解施工安全技术措施的编制要求、主要内容,学习安全事故处理、安全教育与培训。

4.8 施工项目进度管理

工程项目进度管理是指在项目实施过程中,对各阶段的进展程度和项目最终完成的期限所进行的管理。其目的是保证项目能在满足其时间约束条件下实现其总体目标,是保证项目如期完成和合理安排资源供应、节约工程成本的重要措施之一。工程项目进度管理包括工程项目进度计划的制订和工程项目进度计划的控制两大任务。进度控制的措施主要有组织措施、技术措施、合同措施、经济措施和信息管理措施等。

4.9 施工项目质量管理

质量管理是确定质量方针、目标和责任,并通过质量体系中的质量策划、质量控制、质量保证和质量改进,来实现其所有管理职能的全部活动。质量管理的目标是施工项目质量必须由里及外均符合设计要求和合同约定的质量标准,满足建设单位对该项目的功能和使用价值的期望。其主要内容包括质量管理的基础工作、质量体系的设计(策划)、质量管理的组织体制和法规、质量管理的工具和方法、质量抽样检验方法和控制方法、质量成本和质量管理经济效益的评价和计算。

商务英语信函

 请求报价

常用套语

1. We should be pleased to receive your catalogue and price-list.

我们将很高兴收到贵公司的商品目录和价格表。

2. We take great interest in your products and wish to have quotations for the items specified below.

我们对贵公司产品感兴趣,并希望得到以下产品的报价。

3. Would you please send us catalogue, price-list, together with the samples of your products?

您能惠寄贵公司产品的目录、价格表和样品吗?

4. We would ask you to let us have a quotation for…

我们想请贵公司寄一份……的报价单。

5. Please quote your latest prices for the items listed below.

请将下列商品的最新价报到我方。

6. Please kindly quote us your lowest prices for vacuum cleaners.

请告知吸尘器的最低报价。

7. We should be very grateful if you would let us have full details of…

倘若提供优质服务的所有详情,我们将不胜感激。

8. Would you be kind enough to send us the following information as soon as possible?

请尽快告知下列情况。

范文

　　I enclose our new price-list, which will come into effect, from the end of this month. You will see that we have increased our prices on most models. We have, however, refrained from doing so on some models of which we hold large stocks. We feel we should explain why we have increased our prices. We are paying 10% more for our raw materials than we were paying last year. Some of our subcontractors have raised their price by as much as 15%. As you know, we take great pride in our machines and are jealous of the reputation for quality and dependability which we have achieved over the last 40 years. We will not compromise that reputation because of rising costs. We hope, therefore decided to raise the price of some of our machines. We hope you will understand our position and look forward to your orders.

译文

　　现谨附上本公司新价格表,新价格将于本月底生效。除存货充裕的商品外,其余大部分货品均已调升价格。这次调整原因是原材料价格升幅上涨10%,一些承包商的价格调升到15%。过去40年,本公司生产的机器品质优良、性能可靠。今为确保产品质量,唯有稍微调整价格。上述情况,还望考虑。愿能与贵公司保持紧密合作。

被动语态的几种表示形式

　　(1)助动词 be + 及物动词的过去分词构成的短语,例如 pumps are used in every industry,所有的工业都要用泵(副词);动词 + 名词及介词构成的短语;动词 + as + 主语、补语构成的短语,例如:

The plan was approved to…

计划得到了……赞同。(动词 + 介词)

Heat and light are given off by the chemical change.

热和光可由化学反应放出。(动词 + 副词)

Atomic energy can be made use of in the production processes.

原子能是可以在这些生产过程中加以利用的。（动词+名词+介词）

Radio waves are regarded as radiant energy.

无线电波被认为是辐射能。（动词+as+主语补语）

（2）动词 get（或 become 等）+过去分词。例如：

All her works got talked about a little.

对她的所有著作都略加议论一番。

This liquid became mixed with the salt at room temperature.

这种液体在室温下与盐混合了。

（3）用不定式动词，以-ing 结尾的分词或动名词和单个过去分词表示的被动式。例如：

The new type of device seems to have been made.

这种新装置看来已经（被）制成。

Water has the property of dissolving sugar, sugar has the property of being dissolved by water.

水具有溶解糖的性质,而糖具有被水溶解的性质。

Heated, the metal expands.

金属受热就膨胀。

（4）科技英语中常用的被动句型。

专业英语中最常见的被动句型是以 it 为形式主语,that 引导的从句是主语从句的句型。这种句型有含有泛指主语及没有泛指主语之分。现将常用的这种习惯用语句型列述如下：

It is said that…　有人说,据说

It is thought that…　有人认为,据认为

It has been known that…　有人指出

It is well-known that…　大家知道,众所周知

It is considered that…　人们认为,据估计

It is taken that…　有人以为

It is generally accepted that…　人们通常认为

It is noted that…　人们注意

It is supposed that…　人们推测（假定）,据推测（假定）

It is asserted that…　有人主张

It is hoped expected that…　人们希望

It is estimated that…　有人估计,据估计

It is recommended that…　有人推荐

It is stressed that…　人们强调说

It is believed that…　有人相信

It is generally recognized that…　一般认为,普遍认为

It can be seen that…　可以看出,可见

It must be admitted that… 必须承认,老实说

It is stated that… 据称,据说

It is predicted that… 预计

It is understood that… 不用说

It may be safely said that… 可以有把握地说

It is suggested that… 建议

It is demonstrated that… 据证实,已经证明

It is preferred that… 最好

It is decided that… 已经决定

It is arranged that… 已经商定

It cannot be denied that… 无可否认

It has been calculation that… 据计算

另外,专业英语中还常见用 it 作形式主语,后随形容词或名词(名词构成的词组)作表词,that 引导的从句是主语从句的句型,这种句型多用在无主语句中,将常用的这种习惯句型列述如下:

It is clear that… 显然,很明显

It is evident that… 显然,很明显

It is at once apparent that… 是一目了然的

It is apparent that… 很明显

It is obvious that… 很明显

It is noteworthy that… 值得注意的是

It is true that… 的确

It is no wonder that… 无怪乎,不足为奇

It is very likely that… 很可能

It is doubtful that… 值得怀疑

It is possible that… 有可能

It is important that… 重要的是

It is advisable that… 最好

It is preferable that… 最好

It is worth that… 值得

It is worth noting that… 值得注意的是

It is notable that… 值得注意的是

It is necessary that… 有必要,必须

It is desirable that… 要,需要

It is questionable that… 是成问题的

It is essential that… 必须

It is satisfactory that… 令人满意的是

It is nice that… 是太好了

It is probable that… 也许，恐怕
It may be that… 可能
It is enough that… 足以
It is impossible that… 不可能
It is certain that… 无疑，肯定是
It is reasonable that… 是适当的，是合理的
It is appropriate that… 是适当的，是合理的
It seems that… 似乎，看起来
It turned that… 结果是
It happened that… 恰好，碰巧
It remains to be discussed that… 需要讨论
It is a pity that… 可惜，遗憾
It is a mercy that… 幸而
It is of no use that… 是无用的
It is no good that… 是无用的
It is no harm that… 是无害的
It is the case that… 事实是这样的

被动语态的翻译

一、译成汉语被动句

有必要强调被动者或被动动作，或出于语气及修辞需要时，需将英语被动句译成汉语被动句，用汉语中的被动意义用词被、由、让、给、受、是…的、为…所、予以、加以等来表达，例如：

The water is heated by the fuel and is pumped to a boiler.
水被燃料加热后，由水泵压送到热水贮槽。
The electromagnetic disturbances are caused by lightning discharges.
电磁干扰是由雷电放电引起的。
The atomic theory was not accepted until the last century.
原子学说直到上个世纪才被人们所接受。

二、译成汉语主动句

英语被动句译成汉语被动句有下列几种处理方法：
（1）保留原主语及原句结构，翻译时可不用助动词被等。
The energy of motion is called kinetic energy.

运动的能量称作动能。

The transistors are widely used in communication systems.

晶体管广泛用于通信设备中。

(2)将原句中的主语移作宾语,译成无人称句。

No work can be done without energy.

没有能量,就不能做功。

A force is needed to stop a moving body.

要使运动着的物体停下来,需要用力。

(3)将原句中的主语移作宾语,将英语句中的一个适当成分译成汉语中的主语。

Mechanical energy can be changed back into electrical energy through a generator.

发电机能把机械能转变成电能。

(4)将英语句中的主语和谓语合并译成汉语句中的词组。

A brief account has been made about the operation of the new electronic computer.

已经简单地叙述了这台新型电子计算机的操作方法。

Consideration should be given to the machining methods.

应该考虑到机械加工方法。

(5)用 it 代替主语从句的句子,译成不定人称句或无人称句。

It has been found that this machine is similar to the other one in design.

(人们)已经发现,这台机械与那台机械在结构上是相似的。

It should be pointed out that this process is oxidation.

应该指出,这一过程是氧化。

Exercise 1: Match the words and expressions in the left column with the Chinese in the right.

1. construction of hydraulic projects   A. 施工项目管理
2. network planning technique   B. 工程承包合同
3. safety management of construction project   C. 施工项目成本
4. construction project management   D. 水利水电工程建设
5. construction contract   E. 网络计划技术
6. progress management of construction project F. 质量管理
7. quality management   G. 安全事故处理
8. security incident treatment   H. 质量保证
9. quality assurance   I. 施工项目进度管理
10. construction project cost   J. 施工项目安全管理

Exercise 2: According to the content of Text 4, fill in the blanks.

1. According to their own characteristics and composition contents, the basic construction

单元四 水利水电工程施工组织与管理 | 49

projects of water conservancy and hydropower project are divided into _____ types, _____ parts, and _____ projects.

2. The main contents of the contract management are _____ and _____ management, meeting management, _____, default disposal and construction claims.

3. The safety management of construction project grandly includes: safety _____, safety _____, and industrial health.

4. The main measures to control the schedule are _____ measures, technical measures, _____ measures, _____ measures and information management measures, etc..

5. The main contents of network planning optimization include duration optimization, _____ optimization, _____ optimization.

UNIT FIVE　Hydraulic engineering management
水利工程管理

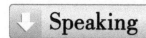

 Planning and scheduling 进度与计划

1. We should work according to the overall schedule chart(the construction time schedule) of the project.

我们应该按照工程项目的总进度表(建设进度表)工作。

2. The effective date of this contract will begin from Dec. 30th, 2007.

这个合同的有效期将从2007年12月30日开始。

3. The seller will provide preliminary(final) technical documents for buyer in May.

卖方将于5月向买方提供初步(最终)技术文件。

4. The basic(detailed) process design will be issued before August.

基本的(详细的)工艺设计资料将于8月前发出。

5. Our major planning items contain estimating of cost and construction schedule.

我们主要的计划工作项目包括费用预算和施工进度。

6. We shall have a design collecting(preliminary design, final design) meeting next month.

我们将在下个月召开设计数据收集(初步设计、最终设计)会议。

7. Field erection work(civil work) will begin in October this year and complete on June 1 next year.

现场安装工作(土建工作)将自今年10月开始至明年6月1日完工。

8. The date of acceptance of this plant will be April sixth, 2008.

这座工厂的交工验收期将在2008年4月6日。

9. The seller's operating group(a crew of specialists) will remain on the job until guarantees are met.

卖方操作组(专业工作组)将在现场一直工作到生产符合保证条件。

10. We must take the plant through the test run and finally into commercial operation.

我们必须使工厂通过试运转并最终投入工业生产。

11. Every month we shall establish construction schedule.

我们每个月都要制订建设进度计划。

12. We shall also make the project schedule report every day.

我们也将每天提出项目进度报告。

13. We are going to begin this work tomorrow(next week, next month).

我们准备明天(下周、下个月)开始这项工作。

14. We must take this work plan into consideration.

我们必须考虑这个工作计划。

15. We have to change our plan for lack of materials(construction machinery, erection tools).

因缺少材料(施工机械、安装工具),我们只能改变计划。

16. What is your suggestion about this schedule?

你对这个进度计划有何建议?

17. Give me your opinion on this plan.

请把你对这个计划的意见告诉我。

Text 5 Hydraulic engineering management

Hydraulic engineering management refers to use, control, schedule hydraulic projects scientifically and rationally, to ensure their safety, normal operation, and full comprehensive benefits. How to strengthen the management of hydraulic engineering to ensure the security and integrity of the project, and how to make full use of the economic benefits will be the future emphasis of hydraulic engineering management. Its work includes check and observation, maintenance and repair, control and application.

5.1 Reservoir control and bank management

Reservoir control, also known as reservoir schedule, is the rational use of existing reservoir to change the distribution of natural runoff of rivers in time and space and water levels in order to meet the needs of production, life and environmental improvement and to achieve the purpose of eliminating pests, promoting benefits, comprehensive using of water resources, including flood control schedule and promoting benefits schedule. A series of hazards will happen on the reservoir banks after reservoir impounding. Therefore, it's necessary to strengthen management and regular inspection, timely prevention and treatment. The usual protective measures generally include construction of dikes, floodwalls, pumping stations, drainage ditches, decompression well and ditch, floodwall embankment, auxiliary dam, bank-protection works, slope reinforcement and other engineering measures. Moreover, some measures are also used such as water quality protection, soil and water loss treatment to protect the water environment of the reservoir bank.

5.2 The safety detection of hydraulic structures

Monitoring namely inspection and observation, refers to supervision, measurement and analysis directly or by special instruments on the foundation and its above hydraulic structures during the entire process from commencement of construction to the first reservoir impoundment and during the operation. According to monitoring objects and the physical factors, the monitoring works during the operational phase include field check and instrument monitoring. The field check is divided into routine check, annual check and special check. The instrument monitoring includes instrument observation and data analysis. The items of instrument observation mainly are observations of deformation, seepage, stress, temperature, and water flow, etc.

5.3 Maintenance and repair of embankment

The various destructions of embankment have a certain development process. Therefore, it is very important to find and treat in time. The daily check should be carried out continuously, and periodic check and special check should also be carried out. The common diseases of embankment include crack, seepage, landslide, slope damage. Crack is divided into the longitudinal, transverse, horizontal, moire-like cracks. Either backfill after excavation or grouting, or both of them is/are often used after the crack is apt to stable. Leakage is divided into the dam body leakage, foundation leakage, contact leakage and leakage around the dam according to the leakage location. The basic processing method is sealing upstream and drainage downstream, that is, providing upstream impervious facilities, and downstream drainage and guide seepage facilities. Landslides are divided into shearing landslide, plastic flow landslide and liquefaction landslide according to their properties. The handling principles are sealing upstream and drainage downstream, upper load reduction and lower weight increasing.

5.4 Maintenance and repair of concrete dam and masonry dam

The diseases of concrete dam and masonry dam mainly include insufficient sliding stability of the dam itself and the foundation, crack and leakage, abrasion damage. Measures to increase the stability against sliding include reducing of uplift pressure, increasing the weight of dam, increasing the coefficient of friction, reducing the horizontal thrust. The treatment methods of the surface crack include surface painting, surface covering, cutting slot and embedded filling, gunite repairing. The treatment methods of inner crack include cement grouting and chemical grouting. The types of seepage are foundation seepage and dam seepage, and their handling principles are upstream truncation and downstream drainage, main truncation and subsidiary drainage. The methods to repair concrete surface are cement mortar repairing, gunite repairing, spraying concrete repairing, vacuum operation of concrete repairing, mud jacking concrete repairing and so on.

5.5 Maintenance and repair of spillway

Because of the insufficient discharge capacity of the spillways in many reservoir projects, the floor of steep slope is lifted by the high velocity flow, side walls are washed off, energy dissipation facilities are destroyed by scouring, the floor of spillway and pier crack and so on, leading to unsafe flood releasing of spillway. The management of the spillway should be strengthened and diseases should be found and eliminated in time. The measures to ensure the safety of the spillway structure can be summarized to four types, truncating seepage, providing a good drainage, resisting the uplift and pulse pressure by the weight of floor, keeping the floor surface smooth.

5.6 Maintenance and repair of sluice

The handling principles of the maintenance and repair of sluice are regular maintenance, timely repair, emphasis on both maintenance and repair. The common problems include resisting sliding stability problem (requiring sufficient weight of barrier to maintain its stability), seepage problem (ensuring effective operation of impervious and drainage facilities when the sluice is working), scouring problem (requiring energy dissipation and scour prevention facilities), settlement problem (ensure the integrity of upper structures of sluice and correct operation of sluice).

5.7 Maintenance and repair of canal structures

Canal structures belong to the canal system supporting structures, and undertake the transporting and distribution of water for irrigation or municipal water supply. Canal structures mainly include tunnel, inverted siphon, culvert, aqueduct, channel and so on. The main diseases include crack leakage, cavitation erosion, abrasion, corrosion of concrete, spalling of concrete surface, steel corrosion, and deposition, etc. The general treatment methods of cracks are to seal the crack surface, fill the crack or join the both side of cracks together. The prevention measures of cavitation erosion are to improve the boundary conditions, control gate opening, improve aeration conditions, use high strength anti-cavitation material and so on. The prevention measures of abrasion damage are to use abrasion resistant materials and so on.

5.8 Maintenance of water resource and hydropower engineering equipment

A number of related equipments are involved in hydropower stations, pumping stations, sluices and other hydraulic projects. The equipments have become the main components of hydraulic projects. According to their constituent and function, the equipments can be divided into metal structure, machinery and electrical equipment. According to the scope and extent of maintenance as well as the scale and the duration of disassembling, the maintenance of equipment can be divided into overhaul, medium, and light maintenance.

5.9 Maintenance and operational management of hydraulic structures

The maintenance and operational management of hydraulic structures includes establishing and improving the rules and regulations, compiling the plan of production and operation, securing the production management. According to the maintenance and operation procedures, the observation items and requirements of hydraulic structures and ancillary equipments should be defined, and observations and inspections should be carried out, including their recording, arrangement and analysis. A planned maintenance and repair management should be executed against the emerging problems according to the safety regulations.

5.10 Flood control and emergency treatment

Flood control means mastering water regime changes and the building conditions in the flood season, mobilizing and strengthening the security and flood control works of hydraulic structures and their downstream buildings, to ensure the safety of reservoir, dike and downstream of the reservoir. The policy of flood control is "safety first, always be on alert, prevention mainly and sparing no effort to rescue". The measures of flood control are mainly divided into engineering measures, including radical and palliative measures, and non-engineering measures, including planning and management of detention basin, forecasting and warning system of flood, emergency evacuation measures, and in-situ evasion measures and so on.

主题 5 水利工程管理

水利工程管理是指对水利工程进行科学合理的运用、控制、调度和保证其安全、正常运行,以充分发挥工程综合效益的工作。如何加强水利工程管理,确保工程的安全和完整,充分发挥工程的经济效益,必将成为今后水利工作的重点。其工作内容主要有检查观测、养护修理和控制运用三个方面。

5.1 水库的控制运用与库岸管理

水库控制运用,又称水库调度,是合理运用现有水库工程改变江河天然径流在时间和空间上的分布状况及水位的高低,以适应生产、生活和改善环境的需要,达到除害、兴利、综合利用水资源的目的。水库调度包括防洪调度和兴利调度。水库蓄水后,常常给库岸带来一系列的危害,所以需加强管理,经常检查,及时防治。库区常用的防护措施一般有修建防护堤、防洪墙、抽水泵站、排水沟渠、减压沟井、防浪墙堤、副坝、护岸、护坡等工程措施,以及针对库岸水环境的保护所采取的水体水质保护、水土流失治理等。

5.2 水工建筑物的安全检测

监测即检查观测,是指直接或借助专设的仪器对基础及其上的水工建筑物从施工开始到水库第一次蓄水整个过程中以及在运行期间所进行的监视量测与分析。运行阶段的监测工作因监测的对象和需测的物理因素不同包括现场检查和仪器监测。其中,现场检查分为日常检查、年度检查和特别检查三种;仪器监测分为仪器观测和资料分析;仪器观测的项目主要有变形观测、渗流观测、应力和温度观测、水流观测等。

5.3　土石坝的养护与修理

土石坝的各种破坏都有一定的发展过程,因此能够及时发现和处理尤为重要,平时要坚持日常检查,并开展定期检查和特别检查。土石坝常见的病害类型主要有裂缝、渗漏、滑坡、护坡破坏。其中,裂缝按走向分为纵向裂缝、横向裂缝、水平裂缝、龟纹状裂缝,在裂缝趋于稳定后采取开挖回填、灌浆和两者结合的处理方法;渗漏按渗漏部位分为坝体渗漏、坝基渗漏、接触渗漏和绕坝渗漏,处理的基本方法是"上堵下排",即在坝的上游设置防渗设施,下游设置排水和导渗设施;滑坡按其性质分为剪切性滑坡、塑流性滑坡和液化性滑坡,其处理的原则是"上堵下排、上部减载、下部压重"。

5.4　混凝土坝及浆砌石坝的养护与修理

混凝土坝及浆砌石坝的病害类型有坝体本身和地基抗滑稳定性不够、裂缝及渗漏、剥蚀破坏。增加其抗滑稳定性的措施有减小扬压力、增加坝体重力、增加摩擦系数、减小水平推力;对表面裂缝的处理方法有表面涂抹、表面贴补、凿槽嵌补、喷浆修补;内部裂缝处理方法有水泥灌浆和化学灌浆;渗漏类型有坝基渗漏和坝体渗漏,处理的基本原则是"上截下排、以截为主、以排为辅";混凝土表层修补的方法有水泥砂浆修补、喷浆修补、喷混凝土修补、混凝土真空作业修补、压浆混凝土修补等。

5.5　溢洪道的养护与修理

许多水库枢纽的溢洪道因其泄流能力不足、陡坡底板被高速水流掀起、边墙被冲毁、消能设施被冲刷破坏、溢洪道底板及闸墩开裂等,溢洪道不能安全泄洪。因此,应加强溢洪道的管理,发现病害应及时处理。高速水流作用下保证溢洪道结构安全的措施归纳为截断渗流、做好排水、利用底板自重压住浮托力和脉动压力、要求底板表面光滑平整。

5.6　水闸的养护与修理

水闸养护修理工作应本着"经常养护、及时修理、养修并重"的原则进行。常出现的问题有抗滑稳定问题(要求水闸必须具有足够的重量以维持自身的稳定)、渗流问题(要求水闸运用中必须确保防渗设施和排水设施的有效工作)、冲刷问题(要求维持消能防冲设施)、沉陷问题(要求必须保证水闸上部结构的完整和正确运用水闸)。

5.7　渠系输水建筑物的养护与修理

渠系建筑物属于渠系配套建筑物,承担灌区或城市供水的输配水任务,主要有隧洞、倒虹吸管、涵管、渡槽、渠道等建筑物。主要病害有裂缝漏水、气蚀、冲磨、混凝土溶蚀、混凝土表面剥落、钢筋锈蚀、淤积等。裂缝的一般处理措施是封闭裂缝表面、充填裂缝或使裂缝两侧结成整体,气蚀的防治措施有改善边界条件、控制闸门开度、改善掺气条件、采用高强度的抗气蚀材料等,冲磨破坏的防治措施主要是采用抗冲磨材料等。

5.8 水利水电工程设备的维护

在水电站、泵站、水闸等水利工程中均涉及一些相关设备,设备已成为水利工程的主要组成部分,按照设备的构成和功能的不同,可将其分为金属结构设备、机械设备和电气设备。根据检修范围与程度以及拆卸的规模和延续的时间将设备检修分为大修、中修、小修。

5.9 水电站建筑物维护和运行管理

水电站建筑物维护和运行管理的工作内容包括建立健全规章制度、编制生产运行计划、安全生产管理。根据维护运行规程明确水工建筑物与附属设备的观测项目及要求,进行观测和检查工作,并做好观测与检查工作的记录、整理和分析,对出现的问题按养护维修安全规程进行有计划的养护维修管理。

5.10 防洪抢险

防洪是指在汛期掌握水情变化和建筑物的状况,做好调动和加强建筑物及其下游的安全防汛工作,以保证水库、堤防和水库下游的安全。防汛方针是"安全第一,常备不懈,以防为主,全力抢险"。防洪的措施主要分为工程性措施和非工程性措施:工程性措施分为治本性的措施和治标性的措施;非工程措施包括滞洪区的规划与管理、洪水预报警报系统和紧急撤退措施、就地避洪措施等。

 预订信函

常用套语

1. I'm writing to make reservation of…
今写此信预订……
2. Please kindly reserve me two seats in…
请代留……座位两个。
3. Could you please make reservation in…for me?
请你们为我在 ……预约行吗?
4. Kindly reserve me two seats in the express, leaving Shanghai for Beijing at 2:30 p.m. the day after tomorrow.
请为我预订两张由上海到北京的特快火车票,时间为后天下午2点30分。

5. Will you please reserve/book a second-class cabin on the … leaving Dalian for Shanghai?

可否为我在大连开往上海方向的……号客轮上订一个二等舱位？

6. I shall thank you if you will reserve me a single room with bath in your hotel for eight nights from July 1st till July 9th. I prefer a room with a view of the sea if possible.

麻烦您替我订一个带浴室的单人房间，时间为7月1~9日。如果可能，窗户最好面临大海。

7. I'll be obliged if you will reserve… for me.

倘能为我预订……，我将不胜感激。

范文

Dear Sir:

　　Our corporation has arranged for a display at the forthcoming World Trade Fair held in London on July 28th, 2004, so we'd like to book two rooms at your hotel for four nights from July 27th to July 31st. Have you got a double room with a bath and a single room with shower?

　　In addition, please kindly reserve us three seats in the plane leaving London for New York at 10:40 a.m. on Saturday, August 1st.

　　Hoping to have an early reply.

Yours faithfully

 否定结构

一、全部否定

1. He is no engineer.

他根本不是工程师。（语气强硬）

2. He is not an engineer.

他不是工程师。（语气一般）

3. Liquids have no definite shape.

液体没有固定的形状。

4. Nowadays it is not difficult to assemble a TV set.

现在组装一台电视机并不难。

句中 not 装饰表语 difficult，译文中否定成分不变。

上面列举的除 no、not 外的其他全部否定词，其中除 never 外，其用法均与 no 相同，亦即：否定语气强硬；英译汉时要进行否定成分的转换，即在原文中否定名词，在译文中要否定动词。Never 的否定语气同 no 一样强硬,但否定成分则和 not 一样,不需进行转换。

例如:

1. None of these metals have conductivity higher than copper.

这些金属的导电率都不及铜高。

2. They have never seen such a device.

他们从来没有见过这样一种装置。

二、部分否定

部分否定指不完全否定,其中有否定因素,也带有一定程度的不否定因素。英语中部分否定通常由 all…not(不都是,不全是)、both…not(不都是,不全是)、every…not(并非每)、not…many(不多)、not…such(一些,不多)、not…often(不常)、not…always(不总是,有时)等构成,例如:

1. All inputs are not 1.

并非所有输入信号都为 1。

该句实际等于 Not all inputs are 1。若照原句字意直译,就译成所有输入信号都不是 1,这就变成全部否定了,因而是错译,如要表示全部否定,则英语句型应该是 None of inputs are 1,所以在翻译时要特别注意。

2. Both the instruments are not precision ones.

这两台仪器并不都是精密的。

3. Every machine here is not produced in that factory.

这里的机械并非每台都是那家工厂制造的。

4. Not much of wasted oil is utilized.

废油被利用的不多。

5. There are not many comrades who agree with me.

同意我的意见的同志不多。

6. The imported TV sets are not always good in quality.

进口电视机并不总是优质的。

值得注意的是,在下列五种情况下,all…not 不能理解为部分否定,实际上构成了全部否定。

1. all+限定词(主要是数词,this 或 these)…+否定式谓语,all 在此句型中可有可无,因而是全部否定。例如:

All nine TV sets are not new, but they are available.

这 9 台电视机都不是新的,但都可用。

2. all(或 both)+肯定式谓语+and(或另加逗号)+否定式谓语,在此句型中 and 之前是全部肯定,and 之后只能是全部否定。例如:

All metals form liquids when melting, and not form solids.

所有金属熔化形成液体,而不形成固体。

3. All +…+ except +…+否定谓语,这种句型中的介词短语 except +…已肯定了

一部分,则其余部分必然是全部否定。例如:

All metals except copper and silver are not good conductors.

所有金属除铜和银外都不是良导体。

4. all(或 both)+…+need not,这种句型为全部否定,这里的 all 或 both 都可有可无,意义不变。例如:

All the parameters, including pressure, temperature, etc. Need not be limited to the presently measured ranges.

各种参数,包括压力、温度等,都不必限制在目前测量范围内。

5. all+…+not+副词,这种句型为全部否定,其中的 not 只否定时间,与 all 无关,例如:

All those things have happened not infrequently.

所有那些事情已不是罕见的了。

三、双重否定

英语中为了语气及叙述需要,有时采用双重否定的结构,双重否定通常由 not…not(no)(没有……没有),…without…not(没有……就不),…not…until(直到……之后才),never…without(每逢……总是),not…but(没有……不……),not(none)…the less(并不……就不),not a little(大大地),not …too…(越……越好),no…other than(不是别的,正是……)等表达。

双重否定在译成汉语时可以保留双重否定,也可译成肯定语气,视译文语气及修辞的需要而定,例如:

1. Without iron and steel we can not develop heavy industry.

没有钢铁,我们就不能发展重工业。(仍译成双重否定)

2. These machines are not to be used without being oiled.

这些机器不上油不准使用。(仍译成双重否定)

3. Aluminum could not be produced on a large scale until the electrolytic process for its reduction had been discovered.

直到发现了还原铝的电解法之后,铝才能大规模地生产。(译成肯定语气)

4. Heat can never be converted into a certain energy without something lost.

热能每逢转换成某种能量时,总是有些损耗。(译成肯定语气)

Exercise 1:Match the words and expressions in the left column with the Chinese in the right.

1. reservoir control and bank management　　A. 水闸的养护与修理
2. the safety detection of hydraulic structures　　B. 水利水电工程设备的维护
3. maintenance and repair of embankment　　C. 水库的控制运用与库岸管理
4. maintenance and repair of concrete dam and
   masonry dam　　D. 水电站建筑物工程管理

5. maintenance and repair of spillway
6. maintenance and repair of sluice
7. maintenance and repair of canal structures
8. maintenance of water resource and hydropower engineering equipment
9. engineering management of hydraulic Structures
10. flood control and emergency treatment

E. 溢洪道的养护与修理
F. 混凝土坝及浆砌石坝的养护与修理
G. 水工建筑物的安全检测
H. 防洪抢险
I. 渠系输水建筑物的养护与修理
J. 土石坝的养护与修理

Exercise 2: According to the content of Text 5, fill in the blanks.

1. According to monitoring objects and the physical factors, the monitoring works during the operational phase include _____ and _____.

2. Measures to increase the stability against sliding include reducing of _____, increasing _____, increasing _____, reducing the horizontal thrust.

3. The handling principles of the maintenance and repair of sluice are _____, timely repair, emphasis on both _____ and _____.

4. A number of related equipments are involved in _____ stations, _____ stations, _____ and other hydraulic projects.

5. A _____ maintenance and _____ management should be executed against the emerging problems according to the safety regulations.

Exercise 3: Translate these sentences.

1. How to strengthen the management of hydraulic engineering to ensure the security and integrity of the project, and how to make full use of the economic benefits will be the future emphasis of hydraulic engineering management.

2. The common diseases of embankment include crack, seepage, landslide, slope damage. Crack is divided into the longitudinal, transverse, horizontal, moire-like cracks.

3. 许多水库枢纽的溢洪道因其泄流能力不足、陡坡底板被高速水流掀起、边墙被冲毁、消能设施被冲刷破坏、溢洪道底板及闸墩开裂等，不能安全泄洪。

UNIT SIX Construction techniques of hydraulic projects
水利水电工程施工技术

Speaking

On construction site 施工现场

1. Welcome to our construction site.

   欢迎来到我们工地。

2. It is very simple and crude here. Do not mind, please.

   这里很简陋,请别介意。

3. Come in, please.　　Be quick.　　Just a minute, please.　　Take care.

   请进。　　　　　　快一点。　　请稍等。　　　　　　　　请小心。

4. I am a site engineer(director, workshop head, chief of section, foreman, worker, staff member).

   我是工地工程师(厂长、车间主任、班组长、领工、工人、职员)。

5. Mr. Wang is responsible for this task.

   王先生负责这项工作任务。

6. Here is our engineering office(drawing office, control room, laboratory, meeting room, common room, rest room).

   这是我们的工程技术办公室(绘图室、调度室、实验室、会议室、座谈室、休息室)。

7. I have some thing(question) to ask of you.

   我想向您请教几个问题。

8. Thanks for your kindness.

   感谢费心!

9. The shift will start at half past seven a.m..

   早班从 7 点半开始。

10. We have flexible work hours during the summer.

我们在夏季的工作时间有弹性。

11. Pay attention to safety!

注意安全!

12. Put on your safety helmet, please.

请戴上安全帽!

13. Danger! Look out! Get out of the way.

危险! 当心! 快躲开!

14. Here is our pipe prefabrication workshop(steel structure fabrication shop, machine shop, boiler room, air compressor station, concrete mixing unit).

这里是我们的管道预制车间(钢结构制作厂、机械加工车间、锅炉房、空气压缩机站、混凝土搅拌装置)。

15. Would you like to see this process(machine)?

你要看看这工艺方法(机器)吗?

16. Would you like to talk to the welder(inspector)?

你要和焊工(检查员)谈谈吗?

17. The factory(work hop, equipment) produces pipe fittings(spare parts, fasteners).

这工厂(车间、设备)生产管件(配件、紧固件)。

18. I am sorry, do not touch this, please.

很抱歉,请勿触动!

19. Smoking and lighting fires are strictly forbidden at here.

这里严禁烟火。

20. Look at the sign, danger keep out.

注意标牌,危险勿进。

21. There is a temporary facility for site brickwork(woodwork, ironwork, paintwork).

这是一个现场砖工(木工、铁工、油工)临时设施。

22. Let me show you around and meet our workers.

让我带你走一圈,并会见我们的工人。

23. We would like to know your opinion about our site work.

我们想听取你对我们现场工作的意见。

24. It is normal.　　It is clear.　　It is correct.　　It is all right.

　　这是正常的。　清楚的。　　正确的。　　良好的。

25. Some training will fit them for the job.

经过一些训练,他们就能胜任这项工作。

26. By the end of this month, we shall have carried out our plan.

到这个月底,我们将已实现我们的计划。

27. All has gone well with our site work plan.

一切均按照我们的现场工作计划进行。

 Engineering material

A(Foreman)　　B(Worker)

A: Has the brake device of the crane been examined?

B: Yes. It has been examined. All right.

A: Be careful, not to break the equipment. You'd better put it down slowly.

B: OK. Don't worry about it.

A: Is the wire rope hoisting safety?

B: There is no problem. There are 5 times safety coefficient.

A: Don't stay under the crane boom, please.

B: I'm sorry. Thank you.

A: OK, let's hoist now.

B: What is to be lifted up?

A: Bricks and cement and concrete.

B: Also those stones?

A: No. These are for the foundation.

B: Also some sand, I think.

A: Will those scaffolds be hoisted or not?

B: Yes, of course. What kind of machine do we still need this afternoon?

A: Machine tools, welders, mixers and vibrators. We still have a lot of things to do, so lift them all up before noon possibly!

B: All right, I will do my best.

 Text 6　Construction techniques of hydraulic projects

With the development of science and technology, the construction technology of hydraulic projects has made great progress. Water conservancy and hydropower project has more and more dependent on the use of technology during the construction process, therefore, the work of construction technique must be done well in engineering construction to ensure the smooth implementation of water resources and hydropower project.

6.1 Blasting engineering

Blasting is to make the soil and rock around explosives broken, thrown or loose, by using the energy released by industrial explosive. So the blasting method is often used to excavate the foundation ditch and needed space of underground building during construction. Some special engineering blasting technologies can also be used to complete some specific construction tasks, such as directional blasting, presplitting blasting, and smooth blasting, etc.

6.2 Masonry project

The materials used in masonry project mainly include brick, stone and cementation materials. Material quality inspection and acceptance should be finished before material approaching. Brick masonry should follow the basic principles: masonry should be built by layer; longitudinal joint between the blocks should be parallel to the direction of the force; longitudinal joint between two upper and lower layers should be staggered.

6.3 Formwork engineering

Template which can make concrete as various prescriptive shapes and sizes meeting the requirements of the design drawings is known as formwork. Formwork and its supporting system compose a formwork system. Formwork can be calcified into planar and curved one according to its shape. The most important in formwork construction is the installation and removal of formwork. The installation of formwork includes lofting, formwork erection, supporting and reinforcement and so on. The order of formwork removal is usually from non-bearing plate to bearing plate and from side plate to bottom plate.

6.4 Steel bar engineering

Steel bar engineering is an important composing part of concrete structure. Steel bar engineering includes purchase, processing, assembling and installation of steel bar and so on. Steel bar infield processing includes rust removal, straightening, cutting off, lengthening, bending and shaping, etc. The cold-forming of steel bar include cold drawing, cold drawn, and cold rolled, etc. Connection of steel bar includes welding, mechanical connection and strapping connection.

6.5 Concrete engineering

The construction process of common concrete is preparation of construction, concrete mixing, concrete transporting, concrete casting, curing of concrete. Special concrete construction technologies include pumping concrete, vacuum treatment concrete, and buried stone concrete construction. Construction of prefabricated concrete component includes construction preparation, placing steel bar, side plate installation, casting and shaping, mold removal and maintenance, product stacking. Construction of prestressed concrete is divided into pretensioned method and post-tensioning method.

6.6 Grouting project

Grouting is to drill and press the slurry with mobility and gelatinization, at a certain ratio requirements, into the cracks of ground or building to cement into a whole for the purpose of seepage control, consolidation, and enhancement. According to its function, grouting can be divided into curtain grouting, consolidation grouting, backfill grouting, contact grouting, and joint grouting, etc. The basic works of grouting construction include drilling, flushing hole, packer permeability test, grouting, sealing hole, and quality inspection.

6.7 Construction diversion and flow control

When hydraulic structures are built on river bed, in order to ensure a dry construction, cofferdam is necessary to maintain foundation ditch to guide the flow to the downstream through predetermined releasing structure. The flow control of water conservancy and hydropower project during the whole construction is to use engineering measures of guiding, cutting-off, blocking, impounding, and releasing to solve the contradiction between construction and water storage and release.

6.8 Foundation treatment

The most important construction aspect is foundation treatment during the construction of water conservancy and hydropower projects. Foundations can be divided into soft ground (including soil and gravel foundation) and the rock foundation according to the nature of the formation. The basic methods for soil foundation treatment include prepressing, piling, replacing, cutoff wall, and curtain grouting, etc. The treatment methods of rock foundation include excavation and concrete backfill, construction of cutoff trench or cutoff wall, anchoring on rock foundation, etc.

6.9 Earth-rock building construction

According to its type, construction of earth and rock works includes excavation, filling and cut and fill, etc. In the construction of water conservancy and hydropower project, there are earth and rock excavation materials, filling materials and other usable materials. The basic principle of the deployment of earth and rock materials is to make full use of materials with temporal match and modest capacity. The construction technologies of embankment include borrow area planning, excavation and transportation of soil material, foundation clearing and foundation treatment, dam filling and compaction.

6.10 Construction of concrete structures

Concrete construction refers to the process of using the mixed concrete to build structures in accordance with engineering design requirements. The main process of ordinary concrete and normal concrete construction include extraction, screening, transportation and storage of concrete aggregate; the design, fabrication, installation and removal of template works; the base surface cleaning; the batching, mixing, transportation, casting, paving, vibrating, treatment of construction joints and maintenance of concrete.

6.11 Tunnel construction

The main works of tunnel construction include the design of inlet and outlet, excavation of cavern and deslagging, lining or supporting of cavern, cavern grouting and quality inspection. In order to ensure a smooth construction, some problems should be properly solved such as power supply of construction, transportation of inside and outside tunnel, ventilation, abatement of smoke and dust, drainage and lighting. The tunnel excavation methods include full face excavation method and pilot excavation method. Tunnel grouting includes backfill grouting and consolidation grouting.

主题6 水利水电工程施工技术

随着科学技术的不断发展,水利水电工程在施工技术上取得了巨大的进步,水利水电工程在施工的过程中越来越多地依赖于技术的运用,因此在工程施工中一定要做好施工技术工作,保证水利水电工程的顺利实施。

6.1 爆破工程

爆破是利用工业炸药爆炸时释放的能量,使炸药周围一定范围内的土石破碎、抛掷或松动。因此,在施工中常用爆破的方式来开挖基坑和地下建筑物所需要的空间,也可以用一些特殊的工程爆破技术来完成某些特定的施工任务,如定向爆破、预裂爆破、光面爆破等。

6.2 砌筑工程

砌筑工程所用材料主要有砖材、石材、胶结材料。材料进场应进行材料质量检验与验收。砖石砌筑应遵循以下基本原则:砌体应分层砌筑,砌块间的纵缝应与作用力方向平行,上、下两层砌块间的纵缝必须互相错开。

6.3 模板工程

能够把混凝土做成符合设计图纸要求的各种规定的形状和尺寸的模子称为模板。模板通常与其支撑体系组成模板系统。模板按形状分为平面模板和曲面模板。模板施工中重要的是模板安装与拆除。模板的安装包括放样、立模、支撑和加固等。模板的拆除顺序一般是先非承重模板后承重模板,先侧板后底板。

6.4 钢筋工程

钢筋工程是混凝土结构的重要组成部分。钢筋工程包括钢筋采购、钢筋加工、钢筋绑扎与安装等。其中,钢筋内场加工包括除锈、调直、下料切断、接长、弯曲成型等工序;钢筋的冷加工有冷拉、冷拔、冷轧等形式。钢筋连接有焊接、机械连接和绑扎连接三种。

6.5 混凝土工程

普通混凝土的施工过程为施工准备→混凝土拌制→混凝土运输→混凝土浇筑→混凝土养护。特殊混凝土的施工工艺有泵送混凝土、真空作业混凝土、埋石混凝土施工。预制混凝土构件施工工序包括施工准备→置放钢筋→安装侧模→浇筑成型→拆模养护→成品堆放。预应力混凝土施工分为先张法和后张法两类。

6.6 灌浆工程

灌浆是通过钻孔,将具有流动性和胶凝性的浆液,按一定配比要求,压入地层或建筑物的缝隙中胶结硬化成整体,达到防渗、固结、增强的工程目的。灌浆按其作用可分为帷幕灌浆、固结灌浆、回填灌浆、接触灌浆、接缝灌浆等。灌浆施工的基本工序为钻孔→洗孔、冲洗→压水试验→灌浆→封孔→质检。

6.7 施工导流与水流控制

在河床上修建水工建筑物时,为保证在干地上施工,需用围堰来维护基坑,并将水流引向预定的泄水建筑物泄向下游。水利水电工程整个施工过程中的水流控制就是指利用"导、截、拦、蓄、泄"等工程措施,来解决施工和水流蓄泄之间的矛盾。

6.8 基础处理

在水利水电工程施工过程中,最重要的施工环节就是工程基础处理工作。地基按地层的性质分为软基(包括土基和砂砾石地基)和岩基。土基处理的基本方法有预压、打桩、置换、做防渗墙、帷幕灌浆等。岩石地基的处理常采用开挖和回填混凝土、修筑截水槽或防渗墙、岩基锚固等。

6.9 土石建筑物施工

土石方工程施工按其工程类型分为挖方、填方及半挖半填等。在水利水电施工中一般有土石方开挖料、填筑料及其他用料,土石方用料调配的基本原则是在进行土石方调配时要做到料尽其用、时间匹配和容量适度。土石坝施工工艺有料场规划、土料的开挖与运输、清基与坝基处理、坝体填筑与压实。

6.10 混凝土建筑物施工

混凝土施工是指按照工程设计要求用拌制好的混凝土修建建筑物的过程。普通混凝土或常态混凝土施工的主要工序有:混凝土骨料的开采、筛分、运输和堆存,模板工程的设计、制作、安装与拆除,基础面清理,混凝土配料、拌制、运输、浇筑、平仓、振捣、施工缝处理和养护。大体积混凝土还有接缝灌浆和混凝土温度控制等工序。

6.11 隧洞施工

隧洞施工的主要工作项目有隧洞的进出口设计、洞室开挖和出渣、洞室衬砌或支护、洞室灌浆以及质量检查等。为保证顺利施工,还必须妥善解决施工动力供应、洞内外交通运输、通风、消烟除尘、排水和照明等问题。隧洞开挖方式有全断面开挖法和导洞开挖法两种。隧洞灌浆有回填灌浆和固结灌浆。

 Writing

◆ 投诉信(Letter making a complant)

投诉信要切记信件不能太长,要一言以蔽之。在此类信函中,要注意写明购物时间,阐明投诉原委,同时要注意措辞恰当,做到有理有据。

参考用语

1. On …(date)I bought from…(place)an instant heater manufactured by your renowned concern.

……(日期)我在……(某地)买了一台由贵厂生产的热水器。

2. I was shocked to find the instant heater purchased on…(date)at…(place)by us did not function well.

我们惊讶地发现,我们……(日期)在……(某地)购买的热水器不太好使用。

3. Your supermarket failed to deliver the furniture I ordered on May 5.

5月5日我们定购的家具你们一直未送货。

4. I am sorry to point out the defect in the oven.

对不起,我得指出烤箱的毛病。

5. The company changed some parts from other manufacturer, but they did not work. Then I write you for further help.

这家公司给我们换上了别的厂家的零件,可还没用。我只得给你们写信请求帮助。

6. The purpose of this letter is ask your permission to send you the bicycle for welding.

我写信的目的是征求你的同意,我们想把自行车送回贵厂焊接。

7. I will appreciate anything you can do to help us on this matter.

这件事情如获得帮助,我将不胜感激。

8. I am aware of your reputation for quality products plus reliability; consequently, I do not hesitate to write to you.

我知道贵公司的产品优良可靠,远近驰名。因此,我毫不迟疑地写上此信。

9. Permit me to thank you for your prompt and considerate reply.

首先多谢您那封又快又周到的回信。

10. I expect the courtesy of a prompt reply from you, and the necessary inspections and corrections from qualified personnel.

盼望你们马上答复,并派遣合格的人员来做出必要的检查与改正。

范文

Dear Sir:

When I bought the motor-cycle from the Anderson Department Store, Main Street, Jonesville, in March of this year, I was told that this was the latest model on the market, and that parts would be easy to obtian.

As you will note, there is a box along with this letter. Inside the box is the broken part, which is the cause of all the trouble. After having talked to some repair shops in my city, I am furious at being told: "The company doesn't have replacement parts" and "It shouldn't be on the market".

I am asking that you either replace the enclosed part, or send me a replacement for that broken one.

Yours sincerely
Mary Jones Warner
15 May ,2006

否定结构

一、形式否定

在英语中,no,not, never, nothing 等否定词与其他适当词搭配使用时,其形式像否定,但实际上含有肯定的意义,形式否定常见的句型有下列三种:

1. can + 否定词 + be + too + 形容词(或 over-过去分词这种结构表示怎么……也不过分或越……越好。例如:

You cannot be too careful in performing experiments.

你做试验时越小心越好。

2. no(not,nothing) + more 或 less + than 这种结构表示不过;仅仅;不少于;至少;无非是,正是。例如:

(1)No more than three kinds of machines have been developed.

我们只研制了三种机器。

(2)No less than 150 devices were wasted.

报废的器件竟多达 150 件。

(3) Semiconductor is nothing more than a material with conductivity between that of conductor insulator.

半导体中只不过是电导率介于导体和绝缘体之间的一种物质。

3. nothing(no, not ,never) + but(without, except)这种句型表示仅是,不过是;只有……过能。例如:

(1)It is nothing but a radio record player.

这不过是一台收录机。

(2)The performance of pump would never be increased except improving the technology.

只有改进工艺,泵的性能才能提高。

二、几乎否定

几乎否定又叫准否定或半否定,表示几乎不,即使有也很少,极少以至于无等含义。这类否定有下列几种类型:

1. 用 hardly(几乎不),scarcely(几乎没有),rarely(很少,难得),seldom(很少,不常),few(很少,几乎没有的),little(很少),unlikely(未必可能的)等词表示,通常与 ever,very 等词连用,以加强语气,例如:

(1)In modern industry,there are hardy any field using no computer.

在现代工业中,几乎没有任何领域不使用计算机。

(2)These phenomena are very unlikely to occur again.

这些现象不大可能再发生。

2. 用含有否定意义的词或表示"少量"意义的词,与 if any , if ever , if at all 等词省略或让步从句连用,如 seldom if ever, little if any, slightly if at all 等表示即使有也很少。例如:

(1)The disturbance of lighting on TV is little if any.

雷电对电视的干扰,即使有也很微弱。

(2)He seldom , if ever , fall ill.

他几乎从不生病。

(3)At normal temperature the substance dissolves only slightly , if at all.

常温下,这种物质即使溶化,也很有限。

3. 用 almost , next , to 等有几乎,近于含义的词与否定词结合,表示几乎不,几乎一个没有。例如:

(1)He has eaten next to nothing.

他几乎什么也没吃。

(2)Almost nobody took any rest.

几乎没有一个人歇过一下。

4. 用否定意义词 + or + 否定词表示极少以至于无,绝对或根本没有,如 little or nothing , seldom or never , little or no , few or no 等,例如:

(1)These fibers for optical communications had little or no attenuation.

这些光纤维几乎没有什么衰减了。

(2)There were few or no noise signals.

噪声信号几乎没有了。

三、意义否定

英语中有些单词或词组,形式上是肯定的,但在意义上却含有否定概念。这类单词或词组,在否定结构中使用广泛,翻译是要多加注意。常见这类单词有:

动词:

fail 不,没有

lack 不够

want 缺乏

neglect 不管

miss 没,不
eliminate 排除
preclude 不用

名词：
lack 没有
failure 不能
departure 不成为
spread 不一致
absence 无

形容词：
reluctant 不愿意
unlikely 未必
insufficient 不足
otherwise 不然

副词：
vainly 不能
less 不多
easily 不难
independently 不依赖于

介词：
without 无,不
beyond 不包括,不在……之内
wanting 如没有
against 不
unlike 不像

连词：
unless 如果不
but 决不是

常见的词组：
but for 如果没有
in the dark 一点也不知道
free from 没有,免于
safe from 免于
short of 缺少
short from 远非,一点也不
in vain 无效,徒然,
too…to(do) 太……以至于不……

but that 要不是
rather than 而不是
instead of 代替,而不是
other than 不是,不同
anything but 决不是
keep(protect, prevent)…form 使不
make light of 不把……当回事

例如:

1. It is the tensile stress which causes the member to fail in its load-carrying function.

正是张应力使构件失去承载作用。

2. In the absence of force, a body will either remain at rest or continue to move with constant speed in a straight line.

没有外力,物体不是保持静止,就是一直沿着直线做匀速运动。

3. This equation is far from being complicated.

这个方程式一点也不复杂。

4. Unless there is motion, there is on work.

没有运动,就没有功。

Exercise 1: Match the words and expressions in the left column with the Chinese in the right.

1. blasting engineering   A. 基础处理
2. masonry project   B. 施工导流与水流控制
3. formwork engineering   C. 土石建筑物施工
4. steel bar engineering   D. 模板工程
5. concrete engineering   E. 爆破工程
6. grouting project   F. 混凝土建筑物施工
7. construction diversion and flow control   G. 钢筋工程
8. foundation treatment   H. 隧洞施工
9. earth-rock building construction I. 砌筑工程
10. construction of concrete structures J. 混凝土工程
11. tunnel construction   K. 灌浆工程

Exercise 2: According to the content of Text 6, fill in the blanks.

1. Blasting is to make the _____ and _____ around explosives broken, thrown or loose, by using the energy released by industrial _____.

2. The most important in formwork construction is the _____ and removal of _____.

3. The construction process of common concrete is preparation of construction, concrete

_____, concrete _____, concrete _____, curing of concrete.

4. When hydraulic structures are built on _____, in order to ensure a dry construction, _____ is necessary to maintain foundation ditch to guide the flow to the _____ through predetermined releasing structure.

5. In the construction of water _____ and _____ project, there are earth and rock excavation materials, filling materials and other usable materials.

Exercise 3: Translate these sentences.

1. Brick masonry should follow the basic principles: masonry should be built by layer; longitudinal joint between the blocks should be parallel to the direction of the force; longitudinal joint between two upper and lower layers should be staggered.

2. Grouting is to drill and press the slurry with mobility and gelatinization, at a certain ratio requirements, into the cracks of ground or building to cement into a whole for the purpose of seepage control, consolidation, and enhancement.

3. 普通混凝土或常态混凝土施工的主要工序有：混凝土骨料的开采、筛分、运输和堆存，模板工程的设计、制作、安装与拆除，基础面清理，混凝土配料、拌制、运输、浇筑、平仓、振捣、施工缝处理和养护。

UNIT SEVEN Drainage system in our country
我国水系

Speaking

💧 Civil engineering 土建工程

1. All of the civil work on the field will be executed by us(our company).
所有现场的土建工作都将由我们(我们公司)承担实施。

2. Civil engineering design is performed on our own at home.
土建工程是根据卖方提供的技术条件设计的。

3. This is a working(plot plan, vertical layout, structure plan, floor plan, general plan)drawing.
这是施工(平面布置、竖向布置、结构平面、屋间平面、总平面)图。

4. The quality of civil engineering conforms to our domestic technical standard(China National Building Code).
土建工程的质量符合我们国内的技术标准(中国建筑法规)。

5. Our civil work includes construction of roads, buildings, foundations and reinforced concrete structure.
我们的土建工作包括建造道路、建筑物、基础和钢筋混凝土结构。

6. It will take two weeks to complete this building(foundation).
这房子(基础)需要两周时间才能完成。

7. Sand, brick, and stone are generally used in constructing houses.
砂、砖和石头通常用于建造房屋。

8. These are the anchor bolts(rivets, unfinished bolts, high-strength structural bolts)for the structure.
这是用于结构的锚定螺栓(铆钉、粗制螺栓、高强度结构用螺栓)。

9. The holes of anchor bolts will be grouted with normal(Portland, non-

shrinkage) cement mortar.

这些地脚螺孔将灌入普通(波特兰、无收缩)水泥砂浆。

10. We usually measure the strength of concrete at 28 days after which has been cast.

我们通常在混凝土灌注后 28 d 测定其强度。

11. The average compressive strength of samples is 500 MPa.

试样的平均抗压强度为 500 MPa。

12. Our concrete material is mixed in a rotating-drum batch mixer at the job site.

我们用的混凝土是在现场的间歇式转筒搅拌机中搅拌的。

13. Quality of concrete depends on proper placing, finishing, and curing.

混凝土的质量取决于适当的灌注、抹光和养护。

14. The workers tend to start the final finishing now.

工人们现在打算开始最后的抹光作业。

15. The concrete can be made stronger by pre-stressing in our factory.

在我们厂里可以使混凝土通过预加应力得到增强。

16. Most of construction material can be tested in our laboratory.

在我们的实验室里可以检验大部分建筑材料。

17. We shall finish the civil work by the end of the year.

在年底前我们将完成土建工作。

Text 7　Drainage system in our country

According to statistics, China's river total length is about 420 000 km, the basin area more than 100 km² of river reaches more than 50 000, the basin area more than 1 000 km² of rivers, there are more than 1 500. These rivers by the resident agencies of the ministry of water resources, namely \" seven big river basin institutions\" , its name and office location are: the Yangtze River Water Resources Commission—organ locus of Wuhan City, Hubei Province; the Yellow River Conservancy Commission, authority is located for the city of Zhengzhou in Henan Province; the Haihe River Water Resources Commission—organ is located for Tianjin City; the Huaihe River Water Resources Commission, authority is located in Bengbu in Anhui Province; Pearl River Water Resources Commission—organ is located for Guangzhou City, Guangdong Province; the Songliao Water

Resources Commission—organ is located for Changchun City of Jilin Province; taihu Basin Authority—organ location is Shanghai.

First river—the Yangtze River in China, the total length of 6 397 km, originated in the Qinghai-Tibet Plateau mount tunggula main—southwest side of Geladaindong sonwy mountain, river flows through Qinghai, Tibet, Yunnan, Sichuan, Chongqing, Hubei, Hunan, Jiangxi, Anhui, Jiangsu, Shanghai and other 11 provinces (autonomous regions and municipalities directly under the central government), into the East China Sea; the Yellow River total length of 5 464 km, is the second longest river in China, originated in the Qinghai-Tibet Plateau Balyanlkalla about Ancient Zong Column Basin at the northern foot of the mountain, through the Qinghai, Sichuan, Gansu, Ningxia, Inner Mongolia, Shanxi, Shaanxi, Henan, Shandong and other nine provinces, autonomous regions, into the Bohai Sea in Shandong Province.

Freshwater lakes in China are mainly distributed in the middle and lower reaches of the Yangtze River and the Huaihe River, such as Poyang Lake, Dongting Lake, Taihu Lake, the Hongze Lake, etc. Poyang Lake is located in the north of Jiangxi Province and on the south bank of the Yangtze River, the lake mainly depends on the surface of the lake surface runoff and precipitation, is China's first big freshwater lake. The Qinghai Lake, Namtso Lake and Hulun Lake are called Salt water Lake, the lakes have the advantage of flood control, water supply aquaculture shipping, tourism and to maintain ecological diversity and other functions. It plays an important role in the whole economic and social sustainable development. International rivers in China are mainly distributed in northeast, northwest and southwest of the three area, such as Heilongjiang, Ussuri River flows through the Russian border; through the border of the Tumen River and Yalu River; the Lancang River, Nujiang River flows through the Sino-burmese border; the border of the Ili River flows; Aksu River flows through the Auspicious border, etc.

主题7 我国水系

据统计,我国河流总长度约为42万km,流域面积超过100 km² 的河流达5万多条,流域面积超过1 000 km² 的河流有1 500多条。这些河流由水利部的派出机构,即"七大流域机构"进行管理,其名称及机关所在地分别为:长江水利委员会——机关所在地为湖北省武汉市;黄河水利委员会——机关所在地为河南省郑州市;海河水利委员会——机关所在地为天津市;淮河水利委员会——机关所在地为安徽省蚌埠市;珠江水利委员会——机关所在地为广东省广州市;松辽水利委员会——机关所在地为吉林省长春市;太湖流域管理局——机关所在地为上海市。

我国的第一大河——长江,全长 6 397 km,发源于青藏高原唐古拉山主峰——各拉丹冬雪山西南侧,干流流经青、藏、滇、川、渝、鄂、湘、赣、皖、苏、沪 11 个省(自治区、直辖市),注入东海;黄河全长 5 464 km,为我国第二大河,发源于青藏高原巴颜喀拉山北麓的约古宗列盆地,流经青、川、甘、宁、蒙、晋、陕、豫、鲁 9 省(自治区),在山东省注入渤海。

我国淡水湖泊主要分布在长江和淮河中下游,如鄱阳湖、洞庭湖、太湖、洪泽湖等。鄱阳湖位于江西省北部、长江的南岸,湖水主要依赖地表径流和湖面降水补给,是我国第一大淡水湖。咸水湖泊有青海湖、纳木错、呼伦湖等。湖泊具有蓄洪、供水、养殖、航运、旅游、维护生态多样性等多种功能,在整个经济社会可持续发展中起着重要作用。

我国国际河流主要分布在东北、西北和西南三大片区,如流经中俄边境的黑龙江、乌苏里江;流经中朝边境的图们江和鸭绿江;流经中缅边境的澜沧江、怒江;流经中哈边境的伊犁河;流经中吉边境的阿克苏河等。

 Writing

 履历表

填写注意事项

履历表(Resume/Curriculum)是对个人学历和工作经历的扼要介绍。它以提纲的形式编写,不必用完整的句子表达,通常包括以下各部分:

(1)个人情况:姓名、性别、出生时间、出生地点、国籍、婚姻状况、健康情况、住址及联系方式。

(2)学历:通常按逆时针顺序排列,一般只写年份。

(3)工作经历:通常按逆时针顺序排列,内容包括单位名称、地址、所任职务。

(4)证明人:填写 1~2 位证明人的姓名、地址(用写信封的格式)、职务。

在书写简历时,一般不用完整的句子,可省去主语,但要用动词过去式,如:

graduated from… specialty(专业)…college in July, 1996.

参考范文

Family Name: Wang First Name: Li
Address: 2nd Floor, 16 Xian St, Dalian
Date of Birth: Dalian
Sex: Female

Marriage:
Married Divorced Single
Educational Records:
1995-1998 Department of Business and Administration, Dalian Commercial Junior College.
1995-1999 Graduated from Dalian High School in July, 1994.
1995-2000 Graduate from Dalian No.9 Middle School in July, 1992.
Employment History: Xingyi Exporting Company.
August 1998-present: 1120 Zhongshan Road, Dalian City.
Secretary to sales manager
Now I apply for a new job. The reason for leaving my present employer is that I am desirous of getting broader experience in trading.

 电子邮件

写作一般包括发件人(From)、收信人(To)、抄送(Cc)、暗送(Blind Cc)、主题(Subject)和正文(Text)等几个部分。其中,抄送(Cc)是指在 Cc 一栏中输入其他人的电子邮件地址就可以将一封信件同时发给多人。暗送(Blind Cc)是指各收件人无法看到同样的信件都给了何人,从而有效地保护了写信人的隐私。主题(Subject)部分便于读信人即刻了解所收信件的内容。因此,主题部分要简洁、明了。电子邮件的正文部分也是由称谓、正文和落款署名三部分组成的。正文部分一般使用非正式的文体,如果正文太长,则可以用附件的方式发送。

电子邮件常用缩写语(Common Acronyms ine-mails)

AFAIK(as far as I know)	据我说知
ASAP(as soon as possible)	尽快
BTW(by the way)	顺便问一下
BFN(bye for now)	再见
CU(see you)	再会
CUL(see you latter)	再联系
DL 或 D/L(download)	下载档案
FAQ(frequent asked question)	常见的问题
FYI(for your information)	供您参考
GA(go ahead)	一直往前
HTH(hope this help)	希望这能帮上忙
IMO(in my opinion)	依我看来
IOW(in other word)	换句话说
JAM(just a moment)	等会儿

O(ocer)　　　　　　　　完毕
OTOH(on the other hand)　　另一方面
POC(piece of cake)　　　小菜一碟
POV(point of view)　　　观点,看法
TKS;TNX(thanks)　　　　谢谢
WRT(with regards to)　　关于

电子邮件范例

From:William Tang〈wtang @ citiz. net〉
To:Bigguy@ yahoo. com
Cc:
Bind Cc:
Subject:Take care
Dear Bigguy,
　　R U busy lately? Haven't seen U for a long time.
　　Take good care of Ur-self, especially at the time of SARS. Remember, the new definition of SARS is "Smile And Remain Smile". Don't be defeated by it.
　　　　　　　　　　　　　　　　　　　　　　　　　　　Yours

同位语和插入语

一、同位语

科技英语中,在某一句子成分后面常用在语法上与之处于同一地位(作同一成分)的另外的词、短语和从句——同位语,对该成分作进一步的说明。同位语与其所说明的句子成分之间的分隔关系有下列几种形式。

(1)同位语紧接在所说明句子成分之后,其后没有逗号。

1. Unlike oxygen, the element nitrogen is not active.
与氧不同,元素氮不活泼。

2. Thanks to the help of the workers we all obtained great success.
由于工人的帮助,我们(大家)都取得了很大的成功。

3. We are all familiar with the three states of matter.
我们都熟悉物质三态。(谓语是 be 时,同位语 all 挪后)

(2)同位语前后有逗号、冒号或破折号。

1. Site investigation provide the engineers with much of the information to evaluate these assumptions, the bases for safe dam design.
坝址的查勘为工程师拟定提供许多资料,这些假定是安全设计的基础。

2. The simplest atom, the hydrogen atom, contains one prot on and one electron.

最简单的原子氢含有一个质子和一个电子。

3. Power may be developed from water by three fundamental processes by action of its weight or of its pressure or of its velocity; or by a combination of any or all three.

利用三种基本方法,即利用水的重力作用、水的压力作用或水的流速作用以及其中任意两种或三种作用的组合,人们可以从中获得动力。

4. Two factors—force and distanced are included in the units of work.

力和距离两个因素都包括在功率的单位内。

(3)同位语用 or 引出。

1. The boiling temperature or boiling point is the temperature at which a liquid boils under ordinary pressure.

沸腾温度即沸点,是液体在常压下沸腾的温度。

2. This distance equals two kilometers or 1.242 8 miles.

这个距离等于 2 km,即 1.242 8 mi。

(4)同位语用 such as 引出。

Some metals, such as copper and silver, are called good conductor.

某些金属,例如铜和银都称为良导体。

(5)同位语用 of 引出,在科技英语中使用这种结构时,of 前常为长度、速度、温度、压力、电阻等度量名称,of 后常为具体的数字或代表数字的符号及国际单位。

1. This automobile is running at a speed of 40 miles an hour.

这部汽车正以 41 mi/h 的速度行驶。

2. This light bulb has a resistance of 120 ohms.

这个灯泡的电阻为 120 Ω。

(6)同位语用连词 that 引出同位语从句。

1. This experiment leads to the conclusion that unlike charges attract each other.

这个实验导致这样一个结论:异性电荷相吸。

2. We are all familiar with the fact that nothing in nature will either start or stop moving of itself.

我们都熟知这样一个事实,即自然界中没有一件东西会自行开始或停止运动。

二、插入语

在科技专业英语中,常用插入语(通常与句子中其他成分语法上的关系)来表示作者对句子所表达的意思的态度。插入语一般用逗号与句子中的其他成分隔开,它可能是一个词、一个短语或一个句子。

(1)常用作插入语的单词:indeed(的确);certainly(当然);surely(无疑的);however(然而);actually, practically(实际上);naturally(自然);obviously(显然)等。例如:

This machine is up to data, indeed.

这台机械的确是最新式的。

（2）常用作插入语的介词短语：in short（简言之）；of course（当然）；in general（一般来说）；in fact（事实上）；in other words（换句话说）；in a word（总之）；in a few words（简言之）；in brief（简言之）；for instance, for example（例如）等。例如：

1. Likewise electronic computer have been programmed to reduce time—consuming mathematical analysis, and, in fact to provide mathematical models.

再则，通过编制程序运用电子计算机来缩短数学分析所耗用的时间，实际上也就是提供一些数字模型。

2. In order to allow for voltage drop in the distribution system, it is customary to rate generator 4% or 5% higher than motor, as for instance, a 2 300-volt generator and 2 200-volt motor.

为了考虑在配电系统中的电压降，习惯上都将发电机的电压率定得比电动机高4% ~ 5%。例如，2 300 V 的发电机配以 2 200 V 的电动机。

（3）常用作插入语的分词短语：generally speaking（一般地说）；strictly speaking（严格地说）；judging from…（从……看来）等。

1. Generally speaking, iron is apt to rust.

一般来说，铁容易生锈。

2. Judging from what you say, he could have done this work still better.

从你说的看来，他满可以把工作做得更好一些。

（4）常用作插入语的不定式短语：so to speak（可说是）；to be sure（无疑地）；to sum up（概括地说）；to tell truth（老实说）；to start with（首先）等。

1. To be sure, the atomic theory helps us a great deal in understanding all matter.

毫无疑问，原子理论大大地帮助我们了解所有的物质。

2. To start with, we must do the test in the following way.

首先，我们必须用下列方法做试验。

（5）常用作插入语的插入词：I am sure（可以肯定地说）；I believe（我相信）；I suppose（我猜想）；that is（也就是说）；as we know it（据我们所知）；as I see it（照我看来）；as it happens（碰巧，偶然）；as it stands（按实际情况来说）；as it were（好像，可以说）等。

1. A compound as we know it results from the chemical union of two or more elements.

据我们所知，化合物是由两个或两个以上的元素化合而成的。

2. A lot of sunlight falls on the earth I think.

我想，有许多阳光落到地球上。

Exercise 1: According to the content of Text 7, fill in the blanks.

1. First rivers—_____ in China, the total length of 6 397 km, _____ in the Qinghai-Tibet Plateau mount Tanggula main—Southwest side of Geladaindong sonwy mountain, river _____ Qinghai, Tibet, Yunnan, Sichuan, Chongqing, Hubei, Hunan, Jiangxi,

Anhui, Jiangsu, Shanghai and other 11 _____ (autonomous regions and municipalities directly under the central government), into the East China Sea.

2. _____ lakes in China are mainly _____ in the middle and lower reaches of the Yangtze River and the Huaihe River, such as Poyang Lake, Dongting Lake, Taihu Lake, the Hongze Lake, etc.

Exercise 2: Translate these sentences.

1. According to statistics, China's river total length is about 420 000 km, the basin area more than 100 km^2 of river reaches more than 50 000, the basin area more than 1 000 km^2 of rivers, there are more than 1 500.

2. 我国国际河流主要分布在东北、西北和西南三大片区,如流经中俄边境的黑龙江、乌苏里江;流经中朝边境的图们江和鸭绿江;流经中缅边境的澜沧江、怒江;流经中哈边境的伊犁河;流经中吉边境的阿克苏河等。

UNIT EIGHT Review test
复习测试

Speaking

💧 Quality control 质量管理

1. Total Quality Control(TQC) is a better quality control system.

全面质量管理(简称TQC)是一种较好的质量管理体系。

2. TQC over the project will be strengthened.

对于这个工程的全面质量管理将要加强。

3. To maintain the best quality of the construction work is the important responsibility of the field controllers.

保持施工工作的优良质量是现场管理人员的重要职责。

4. We possess skilled technician and complete measuring and test instruments used to ensure the quality of engineering.

我们拥有熟练的技术力量和齐全的检测手段,可以确保工程质量。

5. Field inspection work is handled (executed, directed) by our Inspection Section.

现场检查工作由我们的检查科管理(实施、指导)。

6. Our site quality inspector will report to the Project Manager everyday.

我们的现场质量检查员将每天向工程项目经理汇报。

7. I want to see the certificate of quality (certificate of manufacturer, certificate of inspection, certificate of shipment, material certificate, certificate of proof).

我要看看质量证书(制造厂证书、检查证明书、出口许可证书、材料合格证、检验证书)。

8. Here is the report of chemical composition inspection.

这是化学成分检验报告。

9. Is it OK(good, guaranteed, satisfied, passed)?

那是正确的(好的、保证的、满意的、合格的)吗?

10. We shall take the sample to test its physical properties (mechanical properties, tensile strength, yield point, percentage elongation, reduction of area, impact value, Brinell hardness).

我们将取样试验其物理性能(机械性能、抗张强度、屈服点、延伸率、断面收缩率、冲击值、布氏硬度)。

11. We have received Certificate of Authorization for the fabrication and erection of pressure vessels.

我们具有压力容器制作和安装的授权认可证书。

12. The welds passed the examination of radiographic test (ultrasonic inspection, magnetic testing).

这焊缝通过射线透视检查(超声波探伤、磁力探伤)合格。

13. Are you a qualified nondestructive testing(NDT) person?

你是具有资格的无损检测人员吗?

14. Let us go to the laboratory to check the radiographic films.

请到实验室去检查透视片子。

15. This job will have to be done over again.

该项工作必须返工重做。

16. The defect must be repaired at once.

缺陷必须立即修理。

17. This problem of quality needs a further discussion.

这个质量问题需要进一步研讨。

18. The ISO standards have been used by our company in this project.

国际标准(ISO)已被我公司用于此工程中。

19. The testing results fulfill quality requirement.

试验结果达到质量要求。

20. Check list (quality specification) has been signed by the controller (inspector, checker).

检验单(质量说明书)已由管理员(检查员、审核人)签字。

Writing

 促销英文信

促销的目的就是要卖出产品,那么怎样才能把促销信写得吸引人,让人一看就对产品感兴趣呢? 下面就教你促销信的四步写法。

第一步:To arouse attention

促销信都是"不请自来",所以开头一定要有吸引力和诱惑力。It must make an appeal to some particular buying motive and may begin with a question, and instruction, a quotation or an attention—grabing short story. Sometimes if required, it may even begin by suggesting the very opposite to what you want. 看看下面这些开头是怎么写的:

1. Would you like to reduce your rising domestic fuel costs?

2. Why not enjoy the colorful spring by joining the flying Horse Tourist Group after a completely busy winter?

3. Just imagine how comfortable you are when you stretch out those tired limbs on our newly developed White Cloud water bed.

第二步:To create interest and desire

一旦抓住了读者的注意力,就该趁热打铁劝服他们买自己的产品。介绍产品必须紧紧围绕你在信的开头所提出的吸引人之处。光说"最好""最新"是没什么实际意义的,应该强调特性、质量、原材料,以及和同类产品相比最出彩的地方。我们且看下面这个产品介绍是如何写的:

Our recent researches and tests have showed that rooms with our newly developed Energy Savers stay warmer and require 20 percent less fuel than those rooms of the same size without the usage of the savers. The new savers are popular because they are able to store and reflect heat in a much more efficient way. Read the enclosed brochure, you will find that the self-stick backing makes them easy to install yourself.

第三步:To offer conviction

通过产品介绍引起读者的购买欲望后,就该进一步加强读者购买的决心。你可以详细说明并保证产品会给读者带来所承诺的好处。且看下面这段内容是如何吸引客户的:

1. Use our Fast Microwave Oven for two weeks absolutely free.

2. If for any reasons you find the model machine unsuitable to your needs, we will replace

your order or refund you.

第四步：To motivate actions

到了这一步，所有的努力都指向一个目标：促使客户采取行动、购买产品。这时语气要礼貌坚决，并提供给客户如何购买产品的指示，以方便客户购买。下面是一些常用的策略：

1. Why wait? Come and buy right now since a special discount of 15% will only be offered for a month.

2. Don't delay! Those who order by October 5 will receive 100 Oriental Design Christmas cards free.

数字大小及数字增减的表示方法与翻译方法

一、数字大小的表示和翻译

1. more than 50, over 50, about 50

五十多

2. less than 50, under 50, below 50

不到五十，五十以下

They have turned out close to a thousand pieces of pumps.

他们已经生产了近千台水泵。

3. 50 more

再（加）50

First add 25 grams of salt to the water. Then add 50 more grams of salt to it.

先加 25 g 盐于水中，然后再加上 50 g 盐。

4. a mile or more

1 mi 多

a mile or less

不到 1 mi

a long hour

足足 1 h

5. 数字 + add

twenty add, twenty and add

20 多，不到 30

6. (from) twenty to thirty

（从）20 到 30

between 20 to 30

20~30

7. some thirty meters, about thirty meters, thirty meters or so

30 m 左右

a hundred more or less

100 上下

more or less twenty pates

20 页左右

二、数字增减的表示和翻译

英语中所表示的数字增减有两种情况：一种是纯粹数的增加或减少，翻译时按所增、减的数照译；另一种是倍数的增加或减小，翻译时则视情况可能照译，也可能要对数字作适当转换。

1. …+数字(或倍数)+比较级+than…

本句型的数字或倍数多半是净增减的数、净增加倍数或减到 $1/(n+1)$。例如：

(1) X is two more than Y.

X 比 Y 多 2。

(2) Y is two less than X.

Y 比 X 少 2。

(3) A is five time longer than B.

A 比 B 长 5 倍。

(4) Wheel A turns twenty percent faster than wheel B.

A 轮转动比 B 轮快 20%。

(5) C is twice less than D.

C 是 D 的 1/3。

注意：有时仅增加 $n-1$ 倍，故应注意上下的逻辑关系，例如：

Mt. Jolmo Lungma is 8 882 m high, about two and a half times higher than Mt. Fuji.

珠穆朗玛峰高 8 882 m，比富士山高 1 倍左右。

2. …倍数+as+形容词或副词+as+…

本句型中实际上增加 $n-1$ 倍或减少到 $1/n$(或减少 $(n-1)/n$)，例如：

(1) A is twice as long as B.

A 的长度是 B 的 2 倍(或译：A 比 B 长 1 倍)。

(2) W is ten times as light as R.

W(比 R 轻)是 R 重量的 1/10。

3. as+形容词(如 high, many, much 等)+as+具体数字

本句型表示(高，多)达……(具体数字)之意。例如：

The temperature is as high as 6 000 ℃.

温度高达 6 000 ℃。

4. …+as much(many, fast) again as+…

本句型表示净增加倍数。例如：

Wheel A turns as fast again as wheel B.

A 轮转动比 B 快 1 倍。

5. …+by +数字或倍数…

当本句型中有比较级出现或有表示增减意义的动词或分词出现时，by 后表示的都是净增减的数或净增加的倍数。例如：

(1) A is longer than B by 2 inches.

A 比 B 长 2 in。

(2) The prime cost decreased by 60%.

主要成本降低了 60%。

6. 表示增减意义的谓语 + by a factor of + n

本句型只表示增加 n–1 倍或减小 (n–1)/n。例如：

The speed exceed the development will reduce the error probability by a factor of 7.

正在研制的设备将使用误差概率降低 6/7（或译为…降到 1/7）。

7. …倍数（或分数）+数词或名词+…

本句型中实际数值是倍数（或分数）乘上后面的数词所表示的值。例如：

(1) Four earth two is eight.

4×2 得 8。

(2) The earth is 49 times the size of moon.

地球的大小是月球的 49 倍。

(3) The mass of an electron is 1/1 850 that of a hydrogen atom.

电子的质量是氢原子质量的 1/1 850。

(4) The velocity of sound in water is 4 900 ft per second, or more than four times its velocity in air.

水中的声速为 4 900 ft/s，比空气中的声速大 3 倍多（直译则为：……是空气中声速的 4 倍多）。

8. 表示增减意义的动词或分词 + to +数字…

本句型表示增加到（减少到）某个数字。例如：

The member have increased(decreased) to 1 000.

成员增加（减少）到 1 000 名。

9. 增减意义的谓语或词组 + 倍数表示增加 n–1 倍或减少 (n–1)/n。例如：

(1) The sales of industrial electronic products have multiplied six times since 1950.

自 1950 年以来，工业电子产品销售增加了 5 倍。

(2) Switching time of the new-type transistor is shortened 3 times.

新型晶体管的开关时间缩短 2/3（或译为……缩短到 1/3）。

(3) The principal advantage over the old-fashioned machines is a four-reduction in weight.

与老式机器相比的主要优点是重量减少了 3/4。

(4) There is a 50% increase of steel as compared with last year.

钢产量比去年增加了 50%。

10. 用 too 时的数量的增减

(1) It is too long by half.

它过长一半。

(2) He has given me six too few (many).

他少(多)给我六个。

(3) This rod is 3 inches too long.

这杆过长 3 in。

11. 由一些数词组成的词组：a fifty-fifty basis, on a half and half basis (平均地，在对开分的基础上)；in halves (两半)；ten to one (十之八九，很有可能)。

(1) The current flows through T_1 and T_2 on a fifty-fifty basis.

电流平均流过 T_1 和 T_2。

(2) The current has been split in halves.

电流分成两半。

(3) Ten to one, we shall overfulfill our production plan for this year.

十之八九，我们将超额完成今年的生产计划。

12. 英语中 double, treble, quadruple 等动词表示的倍数，都是不包括基数在内，翻译时应把英语中的倍数减少一。例如：

(1) The efficiency of the machines has been more than doubled or trebled.

这些机器的效率已经增加了 1 倍或者 2 倍多。

(2) As the high voltage was abruptly trebled all the valves burnt.

由于高压突然增加了 2 倍，电子管都烧坏了。

As 的主要用法及翻译

一、as 用作关系代词，引出定语从句

1. such…as 像……这(那)样的，像……之类的

在这种结构中，as 和 such 连用，such 通常在句子里作定语用，说明主句里的某个名词，而 as 在定语从句里可能作主语、宾语或表语。例如：

(1) We must know such symbols as are used to represent chemical elements. (as 在从句中作主语)

我们必须知道那些用以表示化学元素的符号。

(2) We hope to get such a tool as he is using. (as 在从句中作宾语)

我们希望能够得到一个像他正在用的那种工具。

(3) Such instruments as (are) thermometers and barometers can be found in any physics laboratory. (as 在从句中作表语)

像温度计和气压计之类的仪器,在任何物理实验室里都能找到。

2. the same…as 和……同样的

(1) If one object is charged with the same kind of electricity as appears on another near by the two objects will repel each other. (as 作从句的主语)

如果一个物体带有它附近的物体上所出现的同种类的电荷,这两个物体就会相互排斥。

(2) This is the same instrument as we stand in need of. (as 作介词 of 的宾语)

这正是我们迫切需要的那种仪表。

(3) He is not the same man as he was. (as 作从句中的表语)

他和过去不同了。

3. as many(as much)…as…(分别用于可数名词和不可数名词)如……一般多;如此之多;凡……的……都

(1) There are as many books as are need. (as 作从句中的主语)

凡是需要的书都有了。

(2) As many instruments as are in the laboratory have been made most use of. (as 作从句的主语)

实验室里那么多的仪器都已充分利用了。

(3) Here is a jar of distilled water, you may use as much as you need. (as 在从句中作宾语)

这里有一瓶蒸馏水,你需要多少就可以用多少。

4. 科技英语文献中常出现由 as 引导的非限制性定语从句,这种非限制性定语从句通常不是用来修饰主句中某一个名词,而是对整个主句所表达的内容作附加说明,它可位于主句之前、之中或之后。

(1) As is often the case we have overfulfilled the production plan.

像往常一样,我们又超额完成了生产计划。

(2) This machine as might be expected has stopped operating.

正如所料,这台机器已停止运转。

(3) He came very early this morning as was usual.

他像往常一样,今天早晨来得很早。

代替整个主句的关系代词 as,常见的句型还有:

(1) as has been said before   如上所述

(2) as(has been) mentioned above(early)   如上(前)所述

(3) as we know   据我们所知

(4) as may be imagined   如可想象出来的那样

(5) as is well-known   众所周知

(6) as we all can see   正如我们大家都能看到的那样

(7) as often happens   正如往常发生的那样

(8) as(it is) shown in   ……如……(图、表等)所示

(9) as will be shown in ……将如……(图、表等)所示

(10) as noted above 如上所述

(11) as is often said 正如通常所说

(12) as has been explained in the preceding paragraph 如上段已解释的那样

(13) as has been already pointed out 正如已指出的那样

二、such as、as such 和 as that 的用法

1. 复数名词，+ such as…，在这种结构中，such as 是复合连接词，引出同位语，以对前面的复数名词起列举作用，可译为例如。

2. …such as，在这种结构中，such as 是代词，作这样的人、事、物解，as 是关系代词，引出定词从句，例如：

(1) This book is not such as I expect. (such 作表语)

这不是一本我所希望的书。

(2) I will explain this law to such as would like to know it.

我将对那些愿意了解这个定律的人解释。

3. as such 意为像这样的人、事、物；作为这样的人、事、物；因而。例如：

(1) We agree to help as such.

我们同意这样的计划。

(2) He is old works, and is respected as such.

他是一位老工人，因而受到尊敬。

4. as that。其中，as 是作者举例用的连词(译为例如)，引出对前面名词起同位语作用的从句。例如：

He told us many pieces of news as that they have set up a university-run factory

他曾告诉我们许多条消息，例如他们已建立了一座校办工厂。

三、as 引出状语从句

单个 as 引出的状语从句有时间从句、原因从句、让步从句、比较从句及方式状语从句，由于在这些从句中只有 as，前后没有其他有关词的搭配衬托，判别从句的语法意义一般需要靠上下文及其逻辑关系来确定。下面分别阐述。

1. as 引出时间状语从句：as 作正当……的时候；随着……解，例如：

(1) As the piston moves down the pressure in the cylinder decreases.

当活塞下移时，汽缸里的压力减小。

(2) The volume varies as the temperature increase.

体积随着温度增加而变化。

2. as 引出原因状语从句：as 作由于、因为解，例如：

Water as it occurs in nature is never very pure.

由于水起源于自然界，从来不是很纯的。

3. as 引出让步状语从句：as 作虽然，尽管，无论解，as 引出的这种让步状语从句的特

点是常用倒装语序,即将从句中的表语、状语或实意动词提前,紧跟着就是 as,然后随其他成分,例如:

(1) Small as atoms are electrons are still smaller.(表语 small 提前)
原子虽然很小,但电子更小。

(2) Much as I should like to see you I am afraid you could not come.(状语 much 提前)
虽然我十分愿意看见你,但恐怕你还是不能来。

(3) Search as they would, they could find nothing in the house.(实意动词 search 提前)
尽管他们在这房子里到处搜寻,但他们找不到任何东西。

4. as 引出比较状语从句

A. 一般比较(从句内容省略与主句相同的成分)

a) as …as 和……一样(同样)的……

(1) Wheel A revolves as fast as wheel B.
A 轮和 B 轮旋转得同样快。(从句中省略了谓语 revolves)

(2) The speed of sound in water is about four times as greet as in air.
声音在水中的速度比在空气中大 3 倍左右。(从句中省略了主语 the speed of sound 及谓语 is)

b) not so …as 不如(或没有)……那样……

(1) Line AB is not so long as line CD.
AB 线不如(没有)CD 线(那样)长。

(2) The melting point of copper is not so high as that of iron.
铜的熔点不比铁的熔点高。

c) not as …as 和……不一样……(强调两者的不同)

(1) The line AB is not as long as the line CD but a little longer.
AB 线和 CD 线的长度不一样,AB 线长一些。

(2) Wheel A does not revolve as fast as Wheel B but much faster.
A 轮和 B 轮的旋转速度不同,A 轮的旋转要快得多。

B. 比拟

a) as…as + 所比拟的事物,其含义为如……一样。

(1) It is as white as snow   白如雪

(2) as quick as lighting   迅如闪电

(3) as hot as fire   热如火

b) as …,(so)…;…as…;…just as…;just as …, so …,其含义为正如……那样。

(1) As two is to three, so is four to six.
4/6 等于 2/3。

(2) Just as water is the most important of liquids , so air is the most important of gases.
空气是气体中最重要的一种,正如水是液体中最重要的一种那样。

c) 用 much as 引出比较从句,表示……和……几乎一样。例如:
We are dealing with something that flows along a conductor, much as water flows through a

pipe.

我们正在研究沿着导体流动的一样东西,这种情况和水通过管道的流动几乎一样。

5. as 和 as if(或 as though)引出方式(状态)状语从句

A. as 按照……样子、方式、办法……

(1) Leave the thing as they are.

让那些东西保持原状。

(2) The letter reads as they are.

那些信的原文如下。

B. as if 和 as though 好像……似的(从句有虚拟语气)

(1) Heat can flow from a hot body to a cooler body as if it were a fluid.

热好像流体一样,能从一个热的物体传到一个较冷的物体。

(2) The molecules of a gas behave as though they were perfectly elastic bodies.

气体分子的性能好像和完全弹性的物体一样。

四、as 引出补语

1. as 引出宾语补语

to refer to … as   把……指为(叫作)
to treat … as   把……当作,以……来对待
to describe … as   把……描述成
to regard …as   把……看成

上列句型中,as 引出的宾语补语有如下几种表示形式。

A. 名词

We usually define energy as the ability to do work.

我们通常下定义说,能是做功的本领。

B. 形容词

We often regard gas as compressible.

我们经常把气体看成是可压缩的。

C. 介词短语

We regard that conclusion as of consequence.

我们把那个结论看成是具有重要意义的。

D. 分词和分词短语

(1) We consider the wire as disconnected.

我们认为这条线没有接好。

(2) We consider this machine as representing the best one in our plant.

我们把这台机器认为是我们厂里最好的一台。

2. as 引出主语补语

将上列的句型由主动句改变为被动态句子后,那么原在主动句中 as 引出的是宾语补语,而在被动句中 as 所引出的补语就变为主语补语了。例如:

主动句:We regard the sun as the chief source of heat and light.
我们认为太阳是主要的热源和光源。
被动句:The sun is regard as the chief source of heat and light.
太阳被认为是主要的热源和光源。

五、as 引出表语或表语从句

1. It may be as you say.
这也许是你说的那样。

2. Thing are not always as they seem to be.
事情并不总是像表面上看来的那样。

六、as 引出同位语和插入语

见单元 7 的"同位语和插入语"部分。

七、常见的固定词组

1. as + 形容词(或副词) + as + 形容词(或副词),意为又……又
(1)This method is as simple as practical.
这种方法又实用又简单。
(2)The wheel turns as fast as stably.
这个轮子转得又快又稳。
2. as…as anything 非常地,无可比拟地
The work is as easy as anything.
这项工作特别容易。
3. as(a)consequence 因此,从而
4. as a consequence of… 由于……的结果
5. as a first approximation 概括地,作为一级近似值
6. as a foundation 作为基础
7. as a general thing 大概,通例
8. as a matter of convenience 为了方便起见
9. as a matter of course 当然,势所必然
10. as a matter of fact 事实上,其实
11. as a matter of record 根据所得的材料(数据)
12. as a means of… 作为……工具(方法)
13. as a result 结果
14. as a result of … 由于……结果,因为
15. as a(general)rule 通常,照例
16. as a whole 整个地,就总体来看
17. as above 如上

18. as against… 与……对照(对比)

19. as…as ever 老是(一贯,常常)

20. as a ever you can 尽你可能

21. as before 依旧

22. as compares with 与……相比

23. as consistent 至于,关于,就……而论

24. as consistent with… 按照,和……相应

25. as consistent to (with)… 与……对比(相反)

26. as distinct from… 和……不同

27. as distinguishes from… 和……不同

28. as early as… 早在

29. as ever 依旧,像往常一样

30. as far as… 远到,到……程度,就……来说

31. as far as…is concerned 就……而论

32. as far as it goes 就现在情况来说

33. as far as possible 尽可能

34. as far as we know 据我们所知

35. as far back as 远在,早在

36. as fast as… 随着;像……那样快

37. as follows 如下

38. as for… 至于,就……而论

39. as good as… 和……一样(好),实际上等于

40. as in the case of 像在……场合下

41. as is also 以及……,还有……也是这样

42. as it (they) did 其实是

43. as it does (they do) 实际上

44. as it is (they are) 放在句首作实际上解,放在句末作照原来样子解

45. as it is seen from… 由……可看出

46. as judged by… 根据……判断

47. as late as… 一直到……

48. as likely as not 或许,多半,说不定

49. as little as may be 越少越好

50. as long ago as… 早在……已经

51. as long as … 只要,……之久

52. as many… 与……数目相同

53. as matters stand 照目前情况来看

54. as much 与前述是相同的

55. as much as to say 等于说

56. as occasion serves 一有机会就
57. as of(1975) 根据(1975)的资料
58. as often happens 如同经常发生的那样
59. as often as not 时常,屡次
60. as opposed to 与……相反(相对比)
61. as per 根据,按照
62. as regards(respects) 关于,至于
63. as(it is)required 根据需要
64. as soon 显然,正如所看到那样

参考文献

[1]高小姣,李晓琳.水利英语[M].上海:复旦大学出版社,2013.
[2]杨登新,李婧.水利英语[M].上海:复旦大学出版社,2015.